Eyes on the Univ
The Story of the Tele

Springer
*London
Berlin
Heidelberg
New York
Barcelona
Budapest
Hong Kong
Milan
Paris
Santa Clara
Singapore
Tokyo*

Eyes on the Universe

The Story of the Telescope

by Patrick Moore

With 78 Figures
including 64 Colour Plates

Springer

Dr Patrick Moore, CBE, FRAS

Cover illustrations: Front cover: The author in front of the 60-inch reflector at Mount Wilson. Back cover (top to bottom): The author with the 24-inch Lowell Refractor at Flagstaff, Arizona; the dome of the William Herschel Telescope, La Palma; a replica of Herschel's Spanish Telescope; an impression of the Hubble Space Telescope.

ISBN 3-540-76164-0 Springer-Verlag Berlin Heidelberg New York

British Library Cataloguing in Publication Data
Moore, Patrick, 1923–
   Eyes on the universe : the story of the telescope
   1.Telescopes – History
   I.Title
   522.2'09
ISBN 3540761640

Library of Congress Cataloging-in-Publication Data
A catalog record for this book is available from the Library of Congress

Apart from any fair dealing for the purposes of research or private study, or criticism or review, as permitted under the Copyright, Designs and Patents Act 1988, this publication may only be reproduced, stored or transmitted, in any form or by any means, with the prior permission in writing of the publishers, or in the case of reprographic reproduction in accordance with the terms of licences issued by the Copyright Licensing Agency. Enquiries concerning reproduction outside those terms should be sent to the publishers.

© Springer-Verlag London Limited 1997
Printed in Great Britain

The use of registered names, trademarks, etc. in this publication does not imply, even in the absence of a specific statement, that such names are exempt from the relevant laws and regulations and therefore free for general use.

The publisher makes no representation, express or implied, with regard to the accuracy of the information contained in this book and cannot accept any legal responsibility or liability for any errors or omissions that may be made.

Typeset by EXPO Holdings, Malaysia
Printed at the University Press, Cambridge
58/3830-543210 Printed on acid-free paper

Contents

Foreword vi

Introduction vii

1   Before the Telescope 1
2   The Very First Telescopes 7
3   Aerial Telescopes – and others 13
4   Enter the Reflector 19
5   William Herschel 26
6   The "Leviathan of Parsonstown" 34
7   The Age of the Great Refractors 43
8   George Ellery Hale and the
    Hooker Telescope 53
9   Palomar 60
10 The New Telescopes 69
11 The Summit of Mauna Kea 82
12 From Sussex to La Palma 88
13 Telescopes of Many Kinds 95
14 Telescopes Beyond the Earth 102

Epilogue   Into the Future 108

Appendix 1   The History of the Telescope 109
Appendix 2   Some Great Telescopes 111

Index 114

Foreword

On 24 April 1957 there appeared, late at night, on BBC Television, a short programme called *The Sky at Night*. It was an idea being tried out for three months to see whether it worked and whether viewers liked it. They did – and every four weeks since then, without a break, Patrick Moore has presented an edition of the programme. It is a remarkable record – even recognized as such in *The Guinness Book of Records*.

Patrick has seen television grow from a single channel operating for only a few hours each evening through the coming of ITV, BBC2 and Channel 4, followed by countless cable and satellite channels. The only programme to keep going without a break through all these changes is – *The Sky at Night*, now celebrating its fortieth anniversary with a programme about the era of the telescope on which this book is based.

Pieter Morpurgo
Producer, BBC TV *The Sky at Night*

Introduction

There can be few people who do not take at least a passing interest in astronomy. Yet there are still many people who have little idea of the make-up of the universe, and there is indeed still some popular confusion between astronomy and the pseudo-science of astrology.

The Sun is a star, no different from many of the stars you can see on any clear night; in fact, astronomers relegate our Sun to the status of a yellow dwarf. Round it move the nine planets, of which the Earth comes third in order of distance. Some of these planets have satellites; we have one – our familiar Moon, which may seem important to us, but is in fact a very minor member of the Sun's family or Solar System. Like the planets, it shines only by reflected sunlight. The star-system or Galaxy to which the Sun belongs includes about 100,000 million stars, together with gas-and-dust clouds which we call nebulae, and a tremendous amount of thinly-spread interstellar material. Beyond, so far away that their light takes many millions of years to reach us, are other galaxies; and the entire universe is expanding, though how far it extends, and how old it is, we do not yet know for certain.

Without telescopes, we would know relatively little about the universe. Optical instruments have been in use for less than five centuries, but in every way they have revolutionized our outlook, and when we came to consider a topic for the 40th anniversary programme of *The Sky at Night* it seemed that the Story of the Telescope was an obvious choice. Hence this book; I hope that you will enjoy reading it.

Patrick Moore
Selsey, 12 February 1997

Acknowledgements

My grateful thanks are due to Pieter Morpurgo, Producer of the *Sky at Night* programme, and to all those who have joined me on the programme over the years; to Paul Doherty, for his line drawings; and in particular to the publishers, above all to John Watson, for help and encouragement.

P.M.

The author with his 15-inch reflector at his observatory at Selsey.

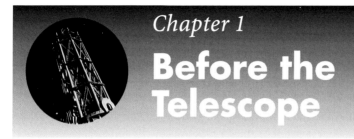

Chapter 1
Before the Telescope

WHEN one thinks about astronomy, one thinks about telescopes. This is natural enough; without their telescopes, astronomers would be very much at a loss. Today we can probe far into the universe, and study objects which are so far away that we see them as they used to be before the Earth or even the Sun came into existence; yet the story of telescopic astronomy extends back for only a few hundred years. Telescopes may have been, and probably were, invented during the sixteenth century, but the first known astronomical observations made with them date back no further than 1609. Previously, all observations had to be carried out with the naked eye alone.

This is not to suggest that ancient civilizations were ignorant of the main astronomical facts. Of course they were not, and all in all it is surprising to see how far they managed to progress. They may have had no real understanding of the make-up of the universe, but they were certainly skilled observers.

It is impossible to say just when astronomy began; presumably it goes back to the far-off times when men lived in caves. Even then they must have looked up into the sky and wondered at what they saw there. The Earth was presumably flat, lying at rest in the centre of the cosmos with everything else moving round it; the Sun and Moon were gods, or else the dwelling-places of gods; everything had been created specially for our benefit.

Among the earliest peoples to make systematic observations were the Mesopotamians, the Egyptians and the Chinese. The first star catalogues were drawn up, and the stars were divided into constellations;

measurements were made, and deductions drawn. For example, it was found that eclipses of the Sun tend to be repeated after a definite cycle, so that they could be predicted – a matter of great importance to the Chinese, who had no idea that an eclipse was caused by the crossing of the Moon over the face of the Sun, and were firmly of the opinion that the Sun was being attacked by a hungry dragon. Significantly, astronomical observations made it possible to draw up useful calendars, and here the Egyptians were particularly concerned. They found that when the brilliant star Sirius could first be seen in the dawn sky each year, the annual flooding of the Nile was imminent, and this was vital to the whole economy of the country.

Some of the earliest data-collectors were the Assyrians, and all students of ancient history know about the Library of Ashurbanipal (BC 668–626). But it was only with the Greeks that astronomy became a true science.

It is not correct to say, as some people do, that the "Greek miracle" happened suddenly. Thales, first of the great philosophers, was born around BC 624; Ptolemy of Alexandria, the last, died around AD 180 – a total span of over eight hundred years, so that in time Thales was as remote from Ptolemy as we are from the Crusades. Knowledge was accumulated gradually. The Earth was found to be a globe rather than a flat plane; there were even a few philosophers, such as Aristarchus of Samos (BC 310–250) who dethroned the Earth from its proud central position and relegated it to the status of a minor world orbiting the Sun. Very accurate star catalogues were compiled, notably by Hipparchus of Nicaea (c. BC 140), and the movements of the naked-eye planets were defined with remarkable precision. Even the phenomenon of precession was discovered. All this work was summarized in Ptolemy's great book which has come down to us by way of its Arab translation as the *Almagest*; it was the most important book of Classical times so far as science was concerned, and it contained a star catalogue which was based on that of Hipparchus but to which Ptolemy had made many contributions. It was also Ptolemy who brought the geocentric or Earth-centred system of the universe to its highest degree of perfection. Of course it was completely wrong, but it did satisfy all the observations which could be made at the time.

Ptolemy's death marked the end of ancient astronomy, and it was not until the time of the Arabs, centuries later, that systematic observations were again made – chiefly, it must be admitted, in the cause of astrology rather than astronomy. The Arab star catalogues were better than those of the Greeks, and there were true observatories equipped with elaborate measuring instruments – notably that of Ulugh Beigh at Samarkand, which was established in 1433 and lasted until Ulugh Beigh himself was murdered in 1449. (He had made the grave mistake of banishing his eldest son on astrological advice, with the result that the errant son returned, led a revolution and had his father killed.) Little remains at Samarkand, but elsewhere many of the pre-telescopic observatories can be seen. For example that at Jaipur, in India, is truly impressive, and is on a vast scale.

Up to the sixteenth century it had been tacitly agreed that the Earth must be the centre of the universe, and the Ptolemaic system was not seriously challenged, partly because of religious prejudice and partly because it seemed to satisfy the observational data. The "great astronomical revolution" began in earnest in 1543, when the Polish cleric and astronomer Mikołaj Kopernik (better known to us as Copernicus) published his book, *De Revolutionibus Orbium Coelestium* (Concerning the Revolutions of the Celestial Bodies) in which he removed the Earth from its central position and replaced it with the Sun.

Remains of Ulugh Beigh's observatory at Samarkand, including the sextant

The observatory at Jaipur, India. Photograph by R.A. Makins.

Basically, he came to this conclusion because he saw that it would remove much of the clumsiness and artificiality of the Ptolemaic system. It had been regarded as axiomatic that all celestial orbits must be circular, because the circle is the "perfect" form and nothing short of perfection could be allowed in the heavens; to account for the movements of the planets in the sky Ptolemy had been obliged to assume that a planet travelled in a small circle or epicycle, the centre of which (the deferent) itself moved round the Earth in a perfect circle.

Yet even this drastic step did not solve the problems, and Copernicus was even reduced to bringing back epicycles. Moreover, tables computed on the basis of the Sun-centred or heliocentric system were no better than their predecessors, and there were also religious considerations to be taken into account; the Church was strongly opposed to any suggestion that the Earth might not be the supreme body of the universe. Tactfully, Copernicus allowed his book to be published only when he was at the very end of his life.

The next character in the story was Tycho Brahe, the Danish nobleman who was unquestionably the greatest of all pre-telescopic observers. In 1572 he saw a brilliant new star in the constellation of Cassiopeia, now known to have been a supernova – a colossal stellar outburst involving the total destruction of a star. Tycho's attention was drawn to the sky, and in 1576 he was able to establish a major observatory on Hven, an island in the Baltic between Malmö and Copenhagen. Here he worked away for twenty years, finally compiling a star catalogue which was a masterpiece of careful, accurate observation. He used large quadrants, and his skill was matched by his patience. He also made measurements of the movements of the planets, particularly Mars. When he died, in 1601, his observations came into the possession of his last assistant, Johannes Kepler, who used them well.

Tycho himself could never believe that the Earth could be in orbit round the Sun; this would go too far against his religious convictions, and instead he supported a hybrid system which satisfied almost nobody. It was left to Kepler to find the true answer. He concentrated upon Tycho's observations of Mars, and found that the data could not be made to fit in with circular orbits of any kind; as he said, the observations agreed "pretty nearly", but not well enough. Finally he realized that the planets move round the Sun in paths

A contemporary painting of Tycho's great quadrant

which are elliptical rather than circular, and between 1608 and 1618 he laid down the three Laws of Planetary Motion upon which all subsequent work has been based.

Note that Kepler was able to make this vital discovery merely by using naked-eye observations. It is ironical that Tycho Brahe died just before telescopes became available – he would have made such good use of them.

This really brings us to the end of the story of pre-telescopic astronomy, and the final proof that the Earth is a planet rather than an all-important central body came from telescopic work. Even so, it was not until near the end of the seventeenth century that the revolution in outlook was complete.

Chapter 2
The Very First Telescopes

IT is generally said that the earliest telescope was made in 1608 by a Dutch spectacle-maker, Hans Lippershey, and that the first man to turn a telescope skyward was Galileo Galilei, two years later. The second of these statements is definitely wrong. Thomas Harriot, one-time tutor to Sir Walter Raleigh, drew a telescopic map of the Moon months before Galileo made his original telescope. It may well be that the first statement is also wrong, and that the first telescope was made in England at some time between 1550 and 1560.

Galileo's telescope was a refractor, in which the light is collected by a special lens known as an *object-glass* or *objective*. The object-glass brings the incoming rays to a focus, where an image is formed and is then enlarged by a second, smaller lens termed an eyepiece. The reflecting telescope works on an entirely different principle. The light is collected by a curved mirror, and

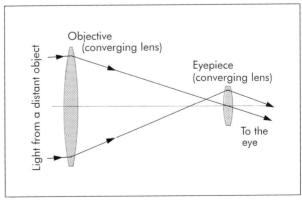

Principle of the astronomical refractor

directed back up the tube on to a smaller flat mirror placed at an angle of 45 degrees; this sends the rays into the side of the tube, where an image is obtained at the focus and enlarged by an eyepiece as before. (This at least was the optical system of the earliest known reflector, made by Isaac Newton.) It may be that the pioneer English telescope was a curious mixture of both types. By no means everyone accepts this view, but evidence collected by C.A. Ronan in the late 1980s does make it sound plausible.

Studies of the reflection of light by mirrors go back a long way; Euclid wrote about it as early as BC 300. There have also been reports of telescopes – of a kind – well before Lippershey. Robert Grosseteste, who wrote a series of books between 1220 and 1235, alludes to a method of "making things a long distance off appear very close", and other hints came in the thirteenth century from Witelo in Eastern Europe, and John Pecham in England. Then there was the "Forerunner", Leonardo da Vinci, who was born in 1452 and died in 1519. He knew a great deal about lenses and mirrors, and his diary contains the significant comment: "Make lenses in order to see the Moon large". It seems that in 1514, when he was working on the Vatican Hill, he was busy on some optical device which was so secret that he would not let even his craftsmen helpers grind the lenses involved in it. Whether Leonardo ever developed any sort of telescope will probably never be known.

So let us turn to Leonard Digges, who was born in or about 1520 and had a decidedly chequered career; at one stage he was unwise enough to take part in a revolution led by Sir Thomas Wyatt against the reigning monarch, Mary Tudor, and was lucky not to be executed. He published mariners' almanacs, and also a best-seller, *Prognostication*, written in English rather than Latin; it was in the form of an almanac-cum-perpetual calendar, with all kinds of scientific facts. Digges died around 1559, leaving one son, Thomas, who was probably born around 1546.

Thomas was educated by Dr John Dee, adviser to Queen Elizabeth, who was a mathematician as well as a mystic and an astrologer. Thomas Digges, like his father, proved to have great mathematical ability. In 1561 he published a book, *Pantometria* (an abbreviation for a very long title) which was mainly about his father's work, and in this he wrote that Leonard had used both lenses and mirrors to make what sounds very like a reflecting telescope. In 1576 he issued a new edition of Leonard's *Prognostication*, and added an appendix of his own supporting the Copernican theory of the universe and arguing that the universe itself was probably infinite. His description of how the stars appear to extend outward until they become too faint to be seen is just what would have been expected from telescopic observations of them.

Another key figure in this strange story was William Bourne, an expert in marine and military affairs. Apparently Lord Burleigh, Queen Elizabeth's chief minister, asked Bourne to report on Digges' claims; after all, a telescope would be of immense use to the nation – remember that the Spanish Armada came in 1588. Bourne did as he was told, and said that with regard to Digges' device "its greatest impediment is that you cannot behold and see but small quantity at a time" – in other words, the instrument had a small field of view.

Bourne's book *Inventions and Devices* gives the best account of the Digges telescope that we know. It needed two optical components, a lens and a mirror. Light entered through the lens, and was refracted on to a curved mirror; there was no eyepiece, so that the observer had to look backwards, so to speak. It was obviously very awkward to use, but in 1992 Ronan and his colleague, G.E. Satterthwaite, built a telescope on this pattern and found that it really worked. (We demonstrated it during a *Sky at Night* television programme in August 1992.)

All this is rather shadowy, and we cannot be sure that Leonard Digges really did build such a telescope – still less that he ever turned it toward the sky. But at least it is possible.

There are various other vague reports of refracting telescopes constructed in the early years of the seventeenth century. For example, a Dutch optician named Zacharias Jansen is said to have made a two-lens combination which magnified distant objects, which he abandoned because the images were upside-down. But for positive evidence we have to wait till the time of Lippershey, who offered his telescope to the authorities in Holland on 2 October 1608; by 1609 telescopes were on sale in Paris, and presumably elsewhere also. They gave erect images, and were used mainly for looking at distant objects such as ships out to sea. Astronomical observations were made by Harriot and others – including an English baronet,

Sir William Lower, who used a telescope to look at the Moon and compared it with a tart that his cook had made: "Here some bright stuffe, there some dark, and so confusedlie all over".

But the great pioneer was of course Galileo. Galileo first heard about the Dutch invention in the summer of 1609, and at once set out to make a telescope for himself. In his book *Sidereus Nuncius* he later wrote that "By sparing neither labour nor expense, I succeeded in constructing for myself an instrument so superior that objects seen through it appeared magnified nearly a thousand times, and more than thirty times nearer than if viewed by the natural powers of sight alone".

During his lifetime Galileo seems to have made about a hundred telescopes, of which a few survive. Each had a convex object-glass and a concave eyepiece, giving an erect image but a very small field. It was Kepler who first introduced a convex eyepiece, which improved the field size and the overall performance at the cost of inverting the image. So far as we know Kepler did not actually make such a telescope, and this was first done around 1615 by Christoph Scheiner at Ingoldstädt, but Kepler's book *Dioptrice* shows that as a theoretical optician he was superior to Galileo. He was not a practical observer (for one thing, his eyesight was weak), and so it was left to Galileo to make the first series of telescopic discoveries.

It did not take him long. He saw the mountains and craters of the Moon, the myriad stars of the Milky Way, the four main satellites of Jupiter, the phases of Venus, and the strange shape of Saturn. Two of these discoveries were of special significance. The fact that Jupiter

Galileo Galilei, the first great telescopic observer

The Very First Telescopes

Galileo's villa, still standing today

had four satellites showed that there was more than one centre of motion in the universe, which was in contradiction to Ptolemy's theory; the four satellites are still known as the Galileans, even though Galileo may not have been the first to see them (he may have been anticipated by a German, Simon Marius). Venus showed a complete set of phases from new to full, and on the geocentric theory this could never happen. It was in fact the behaviour of Venus which gave the final proof that Copernicus must be right and Ptolemy wrong. Galileo also observed sunspots, apparently by using his telescope as a projector rather than looking direct (the story that he damaged his eyesight by staring at the Sun through a telescope seems certainly to be wrong).

Galileo's championship of the heliocentric theory of the universe led him into serious trouble with the Church. The famous story of how he was condemned by the Inquisition, in 1633, and banished to solitary confinement at his villa in Arcetri is not really relevant here, except inasmuch as it was his telescopic work which made him so sure of his facts. (*En passant*, it is interesting to note that it was only in 1992 that Galileo's condemnation for heresy was officially cancelled by the present Pope. Nobody can accuse the Church of making hasty decisions!)

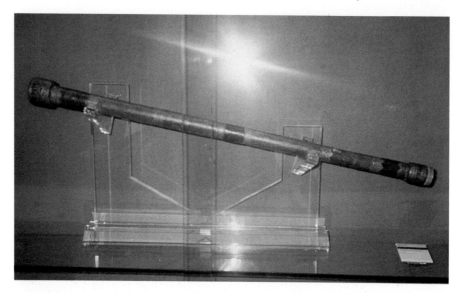

Galileo's telescope, in its display case

Of course Galileo's telescopes were very weak and defective by modern standards, and cannot even compare with present day binoculars, but they ushered in a new era. Look at his original telescope, now on display at the Museum in Florence; it is fascinating to realize that in less than four hundred years this tiny "optick tube" has been developed into the giant telescopes of today.

Chapter 3
Aerial Telescopes – and others

GALILEO'S telescopes may have been low-powered, and by modern standards inefficient, but they were at least manageable. This was emphatically not true of the best refractors of the later seventeenth century, which were so incredibly awkward that one wonders how they can ever have been used successfully.

The main problem was that of false colour – a problem which to some extent is still with us today, though in very reduced form thanks to the development of modern technology. It stems from the fact that light is a wave motion, and what we normally call white light is a mixture of all the colours of the rainbow, from red through orange, yellow, blue and green to violet; red light has the longest wavelength, violet the shortest. If the wavelength is longer than that of red light, it does not affect our eyes; we have in succession infra-red, microwaves and then radio waves. Beyond the short

The unequal refraction of light as it passes through a lens

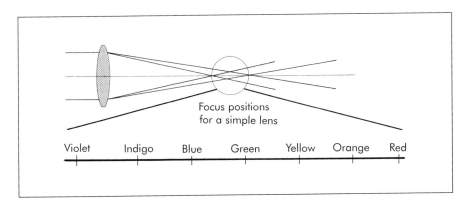

end of the visible range we have ultra-violet, X-rays and then gamma-rays.

Pass a beam of light through a glass lens, and the different parts of it will be bent or refracted by different amounts; red will be refracted the least, violet the most. The result is that a bright object such as a star will be surrounded by gaudy rings which may look beautiful, but which to astronomers are decidedly unwelcome.

The first attempts to counter this effect involved making object-glasses of very long focal length, which certainly reduced the false colour nuisance but which introduced new complications. Telescopes were so long that they were very unwieldy. One observer who used them was Hevelius of Danzig (now Gdańsk) in Poland, who is probably best remembered for his map of the Moon, completed in 1647 and which was a considerable improvement on either Harriot's or Galileo's. (Sadly, the map no longer exists. It was engraved on copper, and after Hevelius' death it is said to have been melted down and made into a teapot.) His longest telescope had a focal length of no less than 150 feet. To provide a monster of this sort with a metal tube would have made it far too heavy, and a paper tube would have been too flimsy, so Hevelius used a sectional arrangement. Each section of the "tube" was made up of two forty-foot wooden planks fixed at right angles to each other so as to form a two-sided trough, and the whole arrangement was braced by wire stays. It was hung from a mast 90 feet high, and operated from below by means of ropes and pulleys. The eyepiece had to be moved by 5 millimetres every 4 seconds to follow the target object across the sky, and since the field of view was very restricted this was in itself quite a problem.

The aerial telescope used by Hevelius in Danzig

Next in line was Christiaan Huygens, of Holland, whose first telescope had a 2-inch object-glass with a focal length of 12 feet. (Huygens was probably the best observer of the time, though he is best remembered today as being the inventor of the pendulum clock.) Encouraged by Hevelius' success, he made a telescope of focal length 23 feet, and it was with this, in 1655, that he discovered Titan, the largest satellite of Saturn. In the following year he studied Saturn itself, and realized that the planet was surrounded by "a thin, flat ring" instead of being a triple body, as Galileo had supposed. This discovery was announced in 1659, and Huygens also made the first sketch to show any recognizable feature on the surface of Mars; his drawing shows the V-shaped feature which we call the Syrtis Major clearly, and since he recorded the date and time of the observation this showed later astronomers that the rotation period of Mars is constant (the modern value is 24 hours 37 minutes 23 seconds). Huygens then progressed to a telescope with a focal length of 150 feet, which seems to have caused great interest among the citizens of his home town. With another telescope, with a focal length of 125 feet, the object-glass was held in a short iron tube, and placed on a high pole; a groove running up and down the pole made it possible for the eyepiece to be raised or lowered as desired, and the image produced by the lens was received in an eyepiece supported by a wooden stand attached to the free end of the thread linking it to the object-glass. The eyepiece itself was hand-held, which must have added to the operating difficulties. On a dark night the observer shone a light on to the object-glass, and then searched around with the eyepiece until he could see the reflection – after which he could align the whole optical system.

Clearly there was a limit to the size of this sort of aerial telescope. The French astronomer Adrienne Auzout, who originally put forward the idea of setting up a national observatory in Paris, planned a 600-foot telescope and even made the eyepiece for it, but it was never built, and would have been well-nigh impossible to use.

However, the Paris observatory was duly established, and was ready by 1667 (it was the first of all national observatories apart from that in Copenhagen, which was opened as early as 1637 but was subsequently destroyed by fire). The first director at Paris was G.D. Cassini, an Italian who had already made his name as a

Huygens' telescope had a focal length of 125 feet

very keen-sighted and accurate observer; it was he who discovered the main division in Saturn's ring system, still known as the Cassini Division, as well as four of Saturn's satellites (Iapetus, Rhea, Dione and Tethys). He too used aerial telescopes, with objectives made by the well-known optician Campani. When Cassini arrived in Paris, at the invitation of the French King, he was faced with another problem. The Observatory building had been designed for appearance, not practical use, and inconvenient turrets meant that the view was hopelessly restricted. In the end Cassini was reduced to taking his telescopes into the Observatory grounds and working from there!

Aerial telescopes were still in use well into the following century. Using one, Francesco Bianchini in 1727 drew what he claimed to be the first map of Venus, even showing what were believed to be oceans and continents, though there can be no doubt that all these features were completely spurious; even a modern telescope will show nothing on this cloud-covered planet, and it was not until the Space Age that we found out what Venus is really like. Earlier, in 1677, Edmond Halley had taken a 24-foot focus telescope to the island of St. Helena to carry out the first systematic survey of the far-southern stars, which can never be seen from the latitude of Britain. But by then it was becoming clear that some sort of solution to the main problem would have to be found.

In 1733 a wealthy amateur astronomer, Chester Moor Hall, hit upon the idea of making a compound object-glass. He fitted together two lenses, one concave and made of flint glass, and the other convex, made of crown glass. These two types of glass refract different colours in different ways, and the two errors tend to cancel each other out, producing a colour-free or achromatic objective. The false colour nuisance can never be completely eliminated, but it can at least be drastically reduced.

Moor Hall, a modest and retiring country barrister, had no wish for publicity, but in the 1750s the matter was taken up by an optical worker named John Dollond. His theoretical understanding was better than Moor Hall's, and after his death, in 1761, his son Peter founded what was probably the first "optical firm" in history. In 1765 he took one of his achromatic objectives to the Royal Observatory at Greenwich; its performance was compared with that of the best long-

The old Paris observatory, at the time of Cassini

focus telescope at the Observatory, and the superiority of the Dollond lens was beyond doubt. The old aerial telescopes promptly became obsolete.

Moreover, Dollond's telescopes looked attractive; from 1783 they were made with brass drawtubes, which were far better than the older paper-covered vellum tubes. When set up on well-made mahogany stands, they were easy to use, and were fitted with slow motions.

But we are running ahead of our story. Even before Dollond's telescopes became widely known, a serious rival to the refractor had been developed. This was the reflecting telescope.

Chapter 4
Enter the Reflector

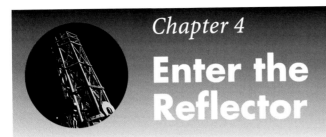

THERE is sometimes a tendency to regard Isaac Newton as infallible. That he was one of the greatest scientists of all time is not in doubt, but he could make mistakes, and one of these concerned the refracting telescope. He could see no way of overcoming the false colour problem, so that in his view the refractor could never be developed to any great extent. Therefore, he adopted a completely different system. Why not dispense with the object-glass altogether, and collect the light by using a curved mirror? Mirrors reflect all wavelengths equally, so that the false colour nuisance would not arise at all (except in the eyepiece, where the effect would be very minor).

Actually, the first idea for a reflecting telescope did not come from Newton, but from a Scottish scientist, James Gregory, in 1663. Gregory proposed to collect the

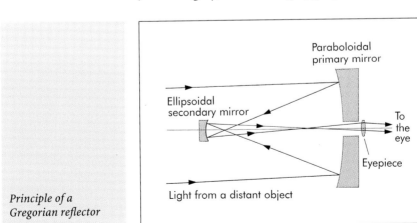

Principle of a Gregorian reflector

Newton's telescope, now in the Royal Society

light by means of a mirror and then reflect the rays back up the open tube on to a smaller, concave secondary mirror. This secondary would send the rays back down the tube once more, and they would pass through a hole in the centre of the main speculum, after which they would be brought to focus and the image enlarged by an eyepiece in the usual way.

Gregory never made such an instrument – it is said that he "lacked practical skill" – and after some unsuccessful experiments he did no more. A few years later, Newton took up the problem. He had already been concerned with investigations into the nature of light, and in 1666, when Cambridge University was closed because of the Plague and Newton was working at his home at Woolsthorpe in Lincolnshire, he had produced the first solar spectrum. Sunlight was passed through a hole in a screen and then through a glass prism, so that the light was spread out into a rainbow strip from red at the long-wave end through orange, yellow, green and blue to violet. It may be said that this was the real birth of the science of spectroscopy.

The pattern of Newton's reflector was different from that of Gregory. After the incoming light had passed down the open tube and been reflected by the main mirror, it was sent on to a flat secondary mirror placed at

Enter the Reflector

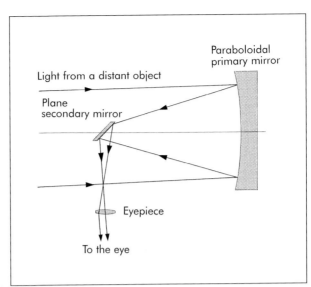

Principle of a Newtonian reflector

an angle of 45 degrees; the rays were then sent to the side of the tube, where they were brought to focus and the resulting image enlarged by an eyepiece. The fact that the secondary was bound to block a little of the incoming light could not be helped, and in fact was not important.

Newton's first reflector may have been made around 1668. In 1671 he presented one to the Royal Society, and it worked well. The main mirror was little more than an inch in diameter, and the whole length of the telescope was less than seven inches, but the principle was absolutely sound, and many modern telescopes, both amateur and professional, are straightforward Newtonians. (My own 15-inch reflector, at my observatory

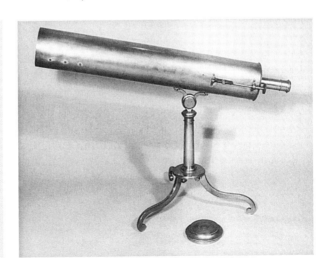

A Gregorian reflector made by James Short

in Sussex, is typical – *see picture on page viii*). The main difference is that the mirrors of today are made of glass, whereas Newton used "speculum metal", an alloy of copper and tin.

The early reflectors did not produce bright images, and another disadvantage was that the mirrors tarnished quickly and had to be regularly re-polished, whereas refractors required practically no maintenance at all. Silver-on-glass mirrors were not to come until the work of Foucault in France, many years after Newton's time.

Another optical system was demonstrated in 1673 by a Frenchman named Cassegrain, about whose personal life virtually nothing is known. His telescope was really a modification of Gregory's; there was a hole in the main mirror, but the concave secondary was replaced by a small convex mirror closer-in to the main speculum. In later years Cassegrains became very popular, and they still are, largely because they are much shorter and more compact than Newtonians and are therefore much easier to handle. Newton himself was not impressed, and wrote: "The advantages of this design are none, but the disadvantages so great and unavoidable that I fear it will never be put into practice." Here again, Newton was wrong.

Meanwhile, there was the question of a national observatory. Britain has always been a seafaring nation, and sailors out of sight of land for long periods were apt to lose their way, sometimes with tragic results. They needed to know two co-ordinates: their latitude and their longitude. Latitude-finding was easy enough; all that had to be done was to measure the height of the Pole Star above the horizon and then make a minor correction to allow for the fact that the Pole Star is not exactly at the celestial pole. (The altitude of the pole is

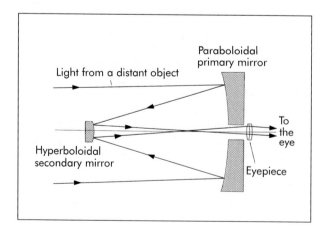

Principle of the Cassegrain reflector

always equal to the observer's latitude on the surface of the Earth.) But longitude-finding was a different matter; it involved accurate timekeeping, and at that time there was no clock which could keep really precise time when on board ship.

The favoured method was to use the rapidly-moving Moon as a clock-hand, and measure its position against the background of stars. This meant using a very accurate star catalogue, and even Tycho's was not good enough, because it had been compiled from naked-eye observations alone. Using telescopic sights would lead to much greater accuracy, as most astronomers recognized (apart from Hevelius, who for some unexplained reason always preferred to keep to the old methods).

Accordingly, that much-maligned monarch King Charles II ordered that a special observatory should be established, and a new star catalogue drawn up for the use of British seamen. Naturally Newton was involved, as were most of the other leading members of the Royal Society of London, which was then – as now – one of the most prestigious learned societies in the world; it had been founded in 1660, soon after Charles' restoration to the throne. Greenwich Park was selected as a site. The Observatory buildings were designed by Christopher Wren, who had been a professor of astronomy at Oxford before turning over to architecture; the Rev. John Flamsteed was put in charge and subsequently given the title of Astronomer Royal, and by 1675 the Observatory was ready. The King had paid for the buildings by the sale of "old and decayed" gunpowder to the French, but Flamsteed was expected to provide his own instruments, which he did. He obtained quadrants and sextants, but he did not need a really large and powerful telescope, because the Observatory was meant to be a navigational establishment rather than a purely astronomical one. So the original telescopes used by Flamsteed were small-aperture refractors, of fairly long focal length but not in the least like the aerial telescopes of Huygens or Hevelius.

Flamsteed did eventually produce the required catalogue, though it took him a long time and the final version did not appear until after his death in 1720. Ironically, the problem of longitude-finding was finally solved when John Harrison, in England, developed the marine chronometer, which was able to keep time accurately even over a long sea voyage. This meant that Flamsteed's catalogue was never used for the purpose for which it had originally been intended.

When Flamsteed died, his widow descended upon the Observatory like an east wind, and removed all the telescopes, which were legally hers. The next Astronomer Royal, Halley, had to begin all over again. What happened to Flamsteed's telescopes we do not know, but it is highly unlikely that any of them survive.

Edmond Halley is best remembered for his connection with the great comet which bears his name, but he was pre-eminent in many branches of science, and as a mathematician was probably inferior only to Newton himself. At Greenwich he undertook a long series of observations of the movements of the Moon, and for this of course he needed telescopes. Some of these can still be seen in the Octagon Room at the old Royal Observatory, designed by Wren. They were used by pointing through open windows; the age of the rotating dome was still far ahead.

Telescope mountings were still of the altazimuth type – that is to say, they could be moved freely either north/south (*altitude*) or east/west (*azimuth*). Later these were superseded by equatorial mounts. With an equatorial, the telescope is set upon an axis which points to the celestial pole, so that when the east/west movement is made the changing altitude looks after itself. (True, the altazimuth mounting has come back into its own in recent years, with the development of computers, but again we are running ahead of our story.) At a fairly early stage it became possible to fit auxiliary equipment to telescopes; for example there was the micrometer, invented by a young Englishman

The Old Royal Observatory at Greenwich

named William Gascoigne. This is an instrument for measuring very small distances and angles. Sadly, Gascoigne was killed at the battle of Marston Moor.

The first really good reflectors were made by John Hadley. In 1721 he produced an excellent 6-inch Newtonian with a focal length of 62 inches, and presented it to the Royal Society. It was compared with Huygens' 125-foot refractor, and found to be just as good in performance as well as being vastly more convenient to use. Hadley made a number of telescopes, mostly Newtonians but also a few Gregorians and Cassegrains, and they became very popular. Then, in the mid-18th century, James Short, in London, began to manufacture telescopes, and produced high-quality Gregorians with mirrors up to 18 inches in diameter. They looked attractive, with their brass tubes and well-designed wooden mountings, and they had also the advantage of producing an erect image, so that they could be used for terrestrial purposes – whereas all other types of astronomical telescopes yielded inverted images. Short was always reluctant to reveal the secrets of his methods, and it is said that before his death in 1768 he destroyed all his notes as well as his manufacturing equipment. This may or may not be true!

By now, telescopes had been developed to the extent of being capable of showing a considerable amount of detail upon the surfaces of the planets; the main emphasis was on the Solar System, since no ordinary telescope, even today, will show a star as anything but a pinpoint of light. However, star-clusters and nebulae also came under scrutiny, and moreover the scale of the universe was at last starting to be appreciated; it had in fact been Cassini who had given the first reasonably good value for the length of the astronomical unit, or distance between the Earth and the Sun. His figure was 86,000,000 miles, which is admittedly rather too small, but is at least of the right order.

If we discount the Digges telescope, of whose real existence we have no firm proof, it is fair to say that the two most significant telescopes of the early period were Galileo's pioneer refractor, and Newton's first reflector. As we have noted, Galileo's telescope is still to be seen in Florence. The telescope at the headquarters of the Royal Society in London is often said to be Newton's first; in fact this is almost certainly not the case, but Newton no doubt had a hand in it. The next great advance came in the latter part of the 18th century, and was due to the genius of one man: William Herschel.

Chapter 5
William Herschel

IT has been claimed that William Herschel was the greatest of all astronomical observers, and this may well be true, but his early career gave no hint of what was to come. He was born Friedrich Wilhelm Herschel, son of a bandmaster in the Hanoverian Guards, on 15 November 1738; Hanover was his home town, and he was educated there. He went to the Garrison School, and soon showed his ability, since mathematics and languages came easily to him. So did music, which was understandable enough. It is strange to recall that his mother, Anna, never learned how to read or write.

At the age of fourteen he joined the Hanoverian Guards as a musician. However, he did not stay long. In 1756 he was sent to England to join the force intended to protect the country from a French invasion; at that time the Seven Years' War was raging, and remember that England and Hanover were united under one king, George II. During this period William took the opportunity to learn English. He was then sent back to Hanover, to serve in the army under the command of the Duke of Cumberland, but in July 1757 the Duke's forces were routed at the Battle of Hastenbeck, and William later recalled how he spent one night in a water-filled ditch. He applied for his discharge from the Army, and this was granted – though in fact he had never been formally enlisted, because he was under age. There was no dishonour in his departure, and the oft-quoted charge that he deserted is quite untrue. He always intended to make music his main career, and he decided to try his luck in England; with his brother Jacob he arrived in London in the autumn of 1757. Apart from a few fleeting visits to the Continent, the rest of his life was spent in England.

He began earning money by copying music, but then he obtained a post as organist at Halifax in Yorkshire; he was undoubtedly a gifted player, and he was an immediate success. In 1766 he was appointed organist at the Octagon Chapel in the fashionable resort of Bath, and his real career began. With his good looks, his pleasant personality and his musical skill he fitted well into the Bath scene, and he was kept busy with concerts and with teaching; he also composed music, which even if not of the first rank was enthusiastically received. In 1772 he brought his sister Caroline over from Hanover to keep house for him, and this too proved to be a good move. Caroline was a good singer, and later became an excellent astronomer, discovering several comets as well as acting as her brother's assistant.

Herschel's interest in astronomy began around 1766, and he hired a small telescope, but it did not satisfy him, and for some years he had no spare time to devote to his new hobby. Then, in 1773, he moved into a house in New King Street, Bath (not No.19, the present Herschel museum), and decided to make a telescope for himself.

Reflector or refractor? Reflectors seemed to be the better bet, so he set up a small workshop in his house and began to experiment with mirrors. He had no helpers apart from Caroline and his brother Alexander, who had joined him from Hanover and was a capable mechanic, but his patience was infinite. Failure followed failure, but at last he produced a good telescope with a mirror 8 inches in diameter and a focal length of $5\frac{1}{2}$ feet. On 4 March 1774 this new telescope was turned toward the Great Nebula in Orion, and Herschel's true career was well and truly launched.

The optical system was conventionally Newtonian. The mirror was of speculum metal, an alloy of copper and tin, since the technique of making silver-coated glass mirrors had not been perfected. Herschel began regular observing, with the ultimate aim of finding out how the stars were arranged in space. This meant carrying out what he termed "reviews of the heavens", and it was during one of these, undertaken with a 6.2-inch reflector with a focal length of 7 feet, that he made his greatest discovery.

He had moved to a house at No.19 New King Street, and from here, in March 1781, he came across something which was definitely not a star. It showed a disk,

A 7-foot reflector made by Herschel. It is now on display at Greenwich

which no star can do, and it moved slowly from one night to another. Herschel believed it to be a comet, which would be interesting but not vitally important, and indeed his initial paper to the Royal Society was entitled *An Account of a Comet*; but before long the object was found to be a new planet, moving far beyond the orbit of Saturn, the outermost of the planets known in ancient times. After a long discussion, it was finally named Uranus, after the first ruler of Olympus.

One of Herschel's 7-foot reflectors is now on view at the Old Royal Observatory in Greenwich. Whether or not this is the instrument actually used to locate Uranus is not certain; it may have been – if not, it is certainly a twin of the telescope used on that famous occasion.

Herschel next made an excellent reflector with a 12-inch mirror and a focal length of 20 feet, and another with the same focal length but a mirror over 18 inches across. He then decided upon a 36-inch, with a focal length of 30 feet. The first step was to cast the metal blank which was to be turned into the main speculum, and this was in itself a major undertaking, particularly since no mirror of such a size had ever been made before.

In the cellar of No.19 a mould was prepared from horse dung, and on 11 August 1781 all was ready. Unfortunately the mould began to crack, and the mirror blank was damaged. A second attempt ended in near-disaster. Again the metal leaked through the mould, and the fiery torrent spread across the floor, shattering the flagstones. In her diary, Caroline recorded that the workers "were obliged to run out at opposite doors, for the stone flooring (which ought to have been taken up) flew about in all directions as high as the ceiling. My poor Brother fell exhausted by heat and exertion on to a heap of brickbats".

Meanwhile, Herschel's fame was spreading, and George III, who had ascended the throne in 1760, was a genuine enthusiast about astronomy (a small observatory was built at Kew, outside London, specially to enable him to observe the 1769 transit of Venus). In 1782 Herschel was invited to the Court, and naturally he went. While in London he paid a visit to Greenwich, and showed one of his telescopes to the astronomers there; it was tested by the observers, including the Astronomer Royal (Nevil Maskelyne), and found to be far better than any at the Observatory. Herschel had

Herschel's workshop at No.19 New King Street, Bath

become the most skilled mirror-maker not only in England, but in the whole of the world.

Everything went well. Herschel was appointed King's Astronomer (not Astronomer Royal), and granted a pension which made it possible for him to give up music as a career and devote his whole time to astronomy. The King was anxious to have him reasonably nearby, so that he could demonstrate the wonders of the sky to his guests, and so Herschel left Bath – with some regret – and eventually settled at Slough, where he remained for the rest of his long life. He continued to build telescopes, and sold them at reasonable profit, though it is rather strange to find that few of them were ever used for serious work. Observational results seem to have been achieved only with the telescopes used by William and Caroline, and with one which he made for the German astronomer Johann Schröter.

Once settled in Slough, at Observatory House, Herschel planned a new telescope. It was to have a 49-inch mirror and a focal length of 40 feet, far outstripping any previous instrument. Making the mirror took some time; the first attempt was a failure, but the second was satisfactory, and it was this mirror which was always used. The sheet-iron tube was so large that when it had been finished, and before the optics were installed, visitors were able to walk through it. On one occasion the King did so, accompanied by the Archbishop of Canterbury. Apparently the Archbishop

Kew Observatory, in 1982

Remain of the tube of Herschel's great 40-foot focal length, 49-inch diameter reflector

stumbled, prompting the King to comment: "Come, my lord bishop, and I will show you the way to Heaven!"

The optical system was not Newtonian. Herschel decided to do away with the secondary mirror altogether, and tilt the main speculum so that the image could be formed directly at the side of the tube-top. This sounds logical, but in fact it has serious disadvantages; it is hard to adjust and it also causes distortions, so that Herschelian telescopes are how hardly ever used (I admit that I have never looked through one). Yet almost as soon as the 40-foot was ready, it was responsible for the discovery of Saturn's two inner satellites, Mimas and Enceladus. It seemed that the great telescope was destined for a brilliant career.

It was not to be. Certainly it was imposing; the tube was mounted on a wooden framework, and the whole structure could be rotated on a circular revolving base. Ladders fifty feet long gave access to the observer's platform at the top of the tube, and the observer communicated with his assistants through speaking tubes. Generally, of course, the chief assistant was Caroline, installed in a hut at the bottom of the structure, but at least two workmen had to be in attendance to move the telescope around, and the whole arrangement was awkward and clumsy.

So the 40-foot never lived up to its early promise, and the only real discoveries made with it were those of Mimas and Enceladus. The telescope took a long time to be made ready for use, and it did not perform even moderately well except when the air was absolutely calm and steady, which in Slough was not often the case. Moreover the mirror tarnished quickly, and keeping it in a fit state for use was a real problem. Almost all Herschel's work was done with the smaller telescopes, and the 40-foot remained largely idle. The last recorded observation made with it was in August

A model of the mounting used for Herschel's 40-foot refractor

1815, when Herschel looked at Saturn and wrote that "the mirror is extremely tarnished". Well before Herschel's death, in 1822, the telescope was dismantled. The tube remained at Observatory House; it was later damaged by a falling tree. What was left stayed *in situ* until Observatory House was demolished in 1961 (I believe I took the last photograph of it there, the day before demolition began) and the surviving part is now on display at the Old Royal Observatory; there too is the great 49-inch mirror.

Another telescope to be seen at Greenwich is the "large 20-foot", with which Herschel carried out much of his work and with which he discovered thousands of double stars, clusters and nebulae, during his attempts to find out the shape of the Galaxy; he finally decided, correctly, that it is a flattened system, so that when we look along the main plane we see the illusion of the Milky Way. In the 1830s William's son John, himself a brilliant observer, took the 20-foot telescope to the Cape of Good Hope to survey the far-southern sky. He set it up at Feldhausen, outside Cape Town; its site is now marked by a monument in the grounds of the Grove Primary School. The 20-foot was never used again after John Herschel returned to England in 1838, but its place in history is assured.

How good were Herschel's telescopes? Optically they were excellent; this we know, because we have been able to test them by modern methods. The smaller instruments were also convenient to use, and all in all it was only the great 40-foot which must be put down as a relative failure.

Herschel had shown that it was possible to build large good mirrors; he always used altazimuth mountings, though well before the end of his career equatorials had been developed. Photography had begun, and indeed John Herschel was a pioneer in this field; he is credited with being the first to introduce the terms "positive" and "negative". Yet it was to be some time before photographic methods could be used together with telescopes, and before then there was another remarkable episode in the history of astronomy: the building of the Leviathan of Parsonstown.

Chapter 6
The "Leviathan of Parsonstown"

WE come now to the most extraordinary telescope in history. It was clumsy and awkward to use; it had only a limited view of the sky; it was never suited to a mechanical drive, and it could never be used for serious photography or spectroscopy. Yet it was by far the most powerful telescope ever built up to that time, and it was used to make fundamental discoveries. Its immense light-grasp meant that it could show objects quite beyond the range of any other existing instrument.

The man responsible was the third Earl of Rosse, an Irish nobleman who was born in 1800. Before the death of his father, the second Earl of Rosse, he was known as

A contemporary painting of the Rosse reflector

Lord Oxmantown, and it was under this name that he produced his first scientific papers. His home was at Birr Castle in Central Ireland, not very far from the town of Athlone, and it was here that he grew up. He went to Trinity College Dublin and then on to Magdalen College in Oxford; in 1822 he graduated with first-class honours in mathematics, and we may suppose that even at that stage in his career he made up his mind that science was to be his main interest. But he felt bound to play a part in public life, and between 1823 and 1834 he sat as a Whig in Parliament, representing King's County. After his marriage, in 1836, to Mary Wilmer-Field of Heaton in Yorkshire, his parents handed over Birr Castle and went to live in Sussex, partly because of the better climate there; nobody can pretend that, so far as climate is concerned, Birr is ideal!

Lord Oxmantown, as he then was, began taking astronomy really seriously from about 1828. At once he realized the need for really large telescopes, and, like Herschel before him; he considered the relative merits of reflectors and refractors; also like Herschel, he settled upon the reflecting type. It is true that by that time refractors had become very fashionable, thanks largely to the superb object-glasses made by the German optician Josef Fraunhofer, but to Lord Oxmantown the reflector offered much greater scope, and he decided to try his hand at mirror-making.

He had no apparatus, no skilled helpers and nobody to advise him, so he was very much "on his own". He enlisted the aid of a carpenter and a blacksmith, plus assorted farm labourers from the Birr estate, and began to make experiments on casting metal mirrors. This involved building a forge. Various alloys were tried, and he finally settled upon an alloy of copper and tin mixed in the proportion of 4 to 1. He cast the first blanks, and then developed the first grinding and polishing machine; this was steam-driven, and the speculum was made to revolve very slowly, while the polishing tool was drawn back and forth by one crank and from side to side by another. The engine developed 2 horse-power, and proved to be perfectly adequate. By 1839 he had made a good 36-inch mirror with a focal length of 30 feet, and mounted it in the same way as Herschel had done for his main telescopes, apart from the fact that he rejected the front-view optical system and reverted to the conventional Newtonian.

The telescope worked well. It was tested by two eminent astronomers, James South from England and

Romney Robinson from Armagh Observatory in what is now Northern Ireland, and both were enthralled. Robinson wrote that the telescope was the best he had ever seen – better even than Herschel's giant. Of course it is impossible to be sure about this, because the careers of the two instruments did not overlap, but Robinson may well have been right.

The 36-inch was only a beginning; to see into the far depths of space needed something even more powerful, and in 1840 Romney Robinson wrote:

> Lord Oxmantown is about to construct a telescope of unequalled dimensions. He intends it to be of 6 feet aperture and 50 feet focus ... His character is an assurance that it will be devoted, in the most unreserved manner, to the service of astronomy, while the energy which could accomplish such a triumph, and the liberality that has placed his discoveries in this difficult art within reach of all, may justly be reckoned among the highest distinctions of Ireland.

Robinson's confidence was justified. On 13 April 1842 the huge 72-inch mirror was successfully cast, and work upon figuring it could begin. By then, Lord Oxmantown had become the third Earl of Rosse.

Two main problems had to be faced. One was optical, the other mechanical. Lord Rosse's experience with the 36-inch had shown him how to deal with the first of these, and there was no need to alter any of his well-tried methods; by then he had made a considerable number of mirrors, all of them excellent. (One Rosse mirror was made for Armagh Observatory.) Neither was there any real difficulty about the flat secondary mirror; Lord Rosse did consider using a prism instead, but in the end did not do so. The question of mounting the telescope came into an entirely different category, and it was here, above all, that Lord Rosse showed his wisdom.

Remember, the 72-inch mirror was twice as large as the 36-inch, and it was also very heavy. To try to mount it on the Herschel pattern would have been sheer folly, and so a totally unconventional design was adopted. The 58-foot tube was to be mounted between two massive stone walls, each 70 feet long and 50 feet high, so that it could be turned only toward that part of the sky which lay near the meridian, or north-south line. This would mean that any particular star could be followed for only a limited period. For an object on the celestial equator, the total viewing time available was about an hour each night; both before and after that period, the object would be hidden behind one or other of the stone walls. This was unfortunate, but not so disastrous as might be thought, because a star is at its highest when on the meridian, and less affected by the unsteadiness of the Earth's atmosphere. Of course Lord Rosse would have liked to have had access to the entire sky, so that he could make his own choice of target without having to rely upon the rotation of the Earth to bring it into view, but he was too sensible to overreach himself.

During 1843 and 1844 work went on apace. When completed, the "observatory", if it could be so called, looked very strange indeed. The base of the actual mounting for the telescope was to be a massive joint of cast-iron, and on it was bolted a cubical wooden box 8 feet wide, carrying the speculum. This box in turn carried the tube which when vertical looked rather like one of the old Irish round towers. The tube itself was 8 feet across in the middle, tapering to 7 feet at either end; it was made from inch-thick deal staves, hooped with iron clamp-rings and strengthened by iron diaphragms. At the top was the secondary mirror, which was so heavy that it had to be counterpoised. The movement in declination was accomplished by means of a strong cable chain fixed to the top of the telescope, passing over a pulley fixed at a suitable height and then down to a windlass on the ground operated by two assistants. The quick movement in right ascension was controlled from below by means of a wheel turned by yet another assistant. For each motion there were fine adjustments, so that the minor movements needed to keep an object in view could be made by the observer who was actually using the telescope. There was no finder; objects were located by using a low-power, wide-field eyepiece, and this must in itself have needed a remarkable degree of skill.

It was all decidedly complex, and moreover it was once said that anyone using the telescope had also to be a trained mountaineer. There is some justification in this, though, to quote Romney Robinson again, "Though it is rather startling to a person who finds himself suspended over a chasm 60 feet deep, without more than a speculative acquaintance with the properties of trussed beams, all is perfectly safe." Certainly there were no mishaps during the whole of the first sixty-year period when the telescope was in operation.

Not unnaturally, the 72-inch was generally known as "the Leviathan of Parsonstown", and this nickname is still often used today.

By 1845 the telescope was ready, but there was a delay in starting a full programme of observation, because this was the time of the potato famine in Ireland, and Lord Rosse was busy in relief work. It was 1848 before observing was really under way. But even before then, the first major discovery had been made.

In 1781 Charles Messier, an energetic French observer, had drawn up a catalogue of star-clusters and nebulae, mainly, it must be added, because he was a comet-hunter and merely compiled his list of nebulae as "objects to avoid". The nebulae themselves appeared to be of two kinds. Some, such as the Great Nebula in the Sword of Orion (No.42 in Messier's list) looked as if they were made up of shining gas; others, such as Messier 31 in the constellation of Andromeda, appeared to be starry. Lord Rosse turned the Leviathan to one of the "starry nebulae", Messier 51 in the constellation of the Hunting Dogs, and found that it was spiral in form, like a Catherine-wheel. Other nebulae also proved to be spiral, though there were also some which were elliptical, circular or else irregular in outline. At that time, of course, the nature of the spirals was unknown; it was only much later that spectroscopic analysis showed that the nebulae of the Orion type were truly gaseous, while the spirals and other "starry nebulae" were independent galaxies, many millions of light-years away.

Once the telescope was in full operation, Lord Rosse began an energetic observing programme, and he

Lord Rosse preparing to observe with the 72-inch reflector at Birr

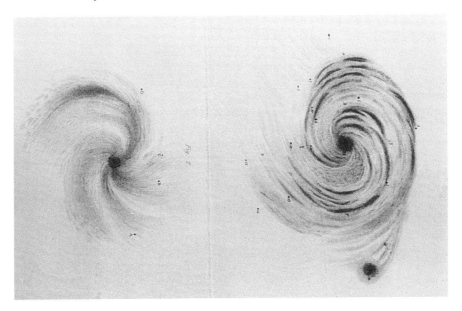

Spiral galaxies, drawn by Lord Rosse

engaged assistants to help in the work, so that very few clear nights were wasted. True, there were continuing problems. It was far from easy to keep the 72-inch mirror in peak condition; it tarnished quickly, particularly in view of the damp climate of Central Ireland, and there were times when the telescope underperformed. This led to the often-heard but completely wrong claim that it was optically poor. In fact it was very good indeed; one need only to look at the Rosse drawings of the spirals, which are uncannily similar to modern photographs.

Of course the nebulae were not the only targets. The Moon came under scrutiny, and in 1852 there was even a suggestion that a large lunar map might be drawn up in collaboration with other observatories, though for various reasons the project was never really started. The Rosse drawings of the planets were as good as any previously made, and there were particularly good representations of Jupiter and Mars.

For the detection of very faint objects the Leviathan was in a class of its own; its light-grasp was far superior to that of any other telescope of the time. Moreover, Romney Robinson's comment about the third Earl of Rosse had been very much to the point. He was concerned solely with making all his results available; everything was published in full, and it is said that nobody who came to him for help or advice ever went away unsatisfied.

It is fair to say that the first two decades of the telescope's career were the most fruitful of all; after that, the 72-inch was overtaken by telescopes of more modern type. In 1865 Lord Rosse had to give up observing because of failing health, and he died two years later. His son, the fourth Earl, continued the work for a time; the 36-inch was modified and put on to a fork mounting with the axis pointing at the pole, so that it became equatorial, and it was also given a rotating head to make observing more convenient. A mechanical drive was fitted, which worked reasonably well. Attempts to add a drive to the 72-inch were much less successful, and it has to be admitted that the mounting of the "Leviathan" was not suited to anything of the sort. No photography was attempted with it, even though photography had come very much to the fore by the 1860s, and only a very limited amount of spectroscopy; it was essentially a visual instrument only.

Work on the nebulae continued, and the main programme – incidentally, the last – was undertaken by J.L.E. Dreyer, a Dane who was assistant at Birr between 1874 and 1878, and who drew up a catalogue of nebular objects which is still used, and is still known as the New General Catalogue or NGC even though it is now well over a century old. With Dreyer's departure, the most important part of the first period of Birr astronomy came to an end. The 72-inch was not as effective as it had once been, and the fourth Earl was by then mainly interested in measuring the tiny quantity of heat sent to us by the Moon; for this he built special equipment, and his results proved to be very accurate. The Leviathan was used less and less, and to make matters even less satisfactory the last assistant, Otto Boeddicker, was largely concerned with drawing an elaborate naked-eye map of the Milky Way, which was admittedly a work of art but which was of no scientific value whatsoever.

Astronomical activity at Birr did not come to a sudden end; it simply petered out. The fourth Earl died in 1909; the 72-inch was dismantled, and the mirror was taken to London for safe keeping, leaving the tube lying in a prone position between the stone walls. In 1916 Boeddicker left Ireland, and astronomy at Birr Castle ceased altogether. The tube of the Leviathan was left where it was, a tribute to the genius of its maker.

The various Irish troubles meant that the Rosses were away from Birr for long periods. The fifth Earl

The "Leviathan of Parsonstown"

The Rosse reflector as it was in 1992

died of war wounds in June 1918; a year later there was a serious fire at the Castle, which was uninhabitable for months. Only in 1925 were members of the Rosse family able to return, and by then the situation so far as the telescopes were concerned was depressing in the extreme. With the 72-inch, the staircase and the supports of the wooden structure, together with the weights of the telescope, were in a sad state of decay, and for reasons of safety had to be taken down. The 36-inch was apparently still intact, but subsequently vanished, and its fate is unknown.

There matters rested for several decades. Now, it is pleasant to record, the situation is very different.

Fortunately, the sixth Earl and Countess took the greatest care to preserve everything that had survived; but for this, most of the story would have been lost forever. Then, in 1968, an exhibition was held in Birr town, in which all the Rosse material was displayed; there was even an old refractor, probably made by the third Earl at a very early stage – I had found it in the Castle cellars, and had managed to restore it. There were the eyepieces and other optical parts of the Leviathan, apart from the main speculum; and there were also the drawings of the spirals and other objects. Inevitably it was suggested that it might be possible to bring the telescope back into use. I have an idea that the initial suggestion came from me; at any rate, it was taken up with maximum energy.

As I write these words (January 1997) restoration is in full swing. The stone walls have been checked, and found to be as good as new. The tube has been repaired; the stairs and other parts of the observatory have been recreated according to the original plans, and a new metal mirror is being made, pending the eventual return of the actual speculum from London.

Of course no attempt is to be made to turn the Leviathan into a research instrument, or to alter the mounting so as to make the whole sky available. Its interest is purely historical, but the whole project has generated tremendous amount of enthusiasm, not only in Ireland but worldwide. "First light" is scheduled for the spring of 1998, some time after this book will appear in print.

The story of Birr astronomy is unique. Alone and unaided, the third Earl made what was by far the most powerful telescope ever constructed, and used it to such good effect that he was able to see further, and more clearly, than anyone before him. Nothing of the sort had ever happened before; nothing of the sort can ever happen again. Meanwhile, it is good to know that the Leviathan of Parsonstown is again to be turned toward the skies, showing the spiral nebulae just as the third Earl saw them a century and a half ago.

The restoration (in 1997) of "The Leviathan of Parsonstown"

Chapter 7
The Age of the Great Refractors

WILLIAM Herschel and Lord Rosse might have put their faith in reflectors, but at that period most other people thought differently, and it was the refractor which held sway. This was due largely to the work of Josef Fraunhofer, in Germany, who proved to be a genius at lens-making; it is doubtful whether his object-glasses have been surpassed even now. One of his refractors, an equatorially-mounted 9-inch, had been sent to the Berlin Observatory, and it played a major rôle in yet another curious episode which is very much part of our story.

As we have noted, Herschel discovered Uranus in 1781 not because he was planet-hunting, but because he was carrying out a systematic "review of the heavens" in order to find out how the stars were distributed in space. Before long it was found that Uranus was wandering away from its predicted path, so that an unknown influence was at work. Two mathematicians, John Couch Adams in England and Urbain Le Verrier in France, independently decided that the perturbations must be due to a new planet moving in an orbit well beyond that of Uranus, and each arrived at a position for it. Adams finished his work first, and sent the results to the Royal Greenwich Observatory, but the then Astronomer Royal, Sir George Airy, took no action; Adams had only recently graduated from Cambridge, and it seems that Airy was loath to interrupt the regular Greenwich programme in order to go on what might well prove to be a wild-goose chase.* Le Verrier finished soon afterwards,

---
* If you want the full details, you will find them in my own book *The Planet Neptune* (John Wiley and Co., 1996). It is an extraordinary story.

and, unlike Adams, published his results. In 1846 a copy of Le Verrier's memoir reached Airy, who belatedly decided to organize a search. But there was no suitable telescope at Greenwich, and so Airy passed the matter over to James Challis, professor of astronomy at Cambridge. Here there was a good 11.7-inch telescope, known as the Northumberland refractor after its donor.

Challis was not particularly confident about the accuracy of Adams' prediction, and he was also preoccupied with other observations; in particular he was following a comet, Biela's, which had caused a major astronomical sensation by breaking in half. Neither did Challis have a really good map of the area concerned, and though he followed Airy's instructions he admitted that he "got over the ground very slowly". In the meantime Le Verrier had sent his results to Berlin, and two of the astronomers there, Johann Galle and Heinrich D'Arrest, asked the permission of the Observatory director, Johann Encke, to use the Fraunhofer 9-inch. (Encke's comment, when presented with the request, was: "Let us oblige the gentleman from Paris!") Galle and D'Arrest identified the planet on the first night of their search; it was of course the world we now call Neptune. Only later did Challis realize that he had recorded the planet twice in the first week of his own search – but he had committed the cardinal sin of failing to compare his observations.

The Northumberland refractor (named after its donor) at Cambridge

The Age of the Great Refractors

Lassell's 24-inch telescope – constructed by an amateur – was used to discover Triton, Neptune's largest satellite. This picture shows a reconstruction of the telescope, now at Liverpool

As soon as Neptune had been identified, other telescopes were turned toward it. One of these was a 24-inch metal-mirror made by William Lassell, a wealthy brewer who had made an excellent reputation as an amateur astronomer. The 24-inch showed Neptune well, and in a short while he discovered Triton, Neptune's large satellite. The 24-inch was dismantled after Lassell's death, but has now been reconstituted, and is in use at the Liverpool Museum. (Indeed, in 1996 I was honoured at being invited to perform the opening ceremony.)

It was the Neptune fiasco which induced Airy to order a large telescope for the Royal Observatory. The Northumberland refractor is still at Cambridge, though the original objective has been replaced by a slightly larger one. The telescope performs well, as I know from personal experience. By the second half of the nineteenth century it had become possible to make really large object-glasses, and in 1862 Thomas Cooke, in England, built the first of the "great refractors", the Newall 25-inch, which still exists and is still in use; it is now at the Athens Observatory in Greece. Around this time, too, the first telescope firms were established. One of them was founded in Dublin by Thomas

Grubb, and it was he who was responsible for a giant refractor, the 26-inch for the Vienna Observatory in Austria. The contract with the Grubb firm was signed in 1875; the total cost of the telescope was £8000, and by 1880 it was ready. It proved to be a great success, and Grubb wrote to Sir David Gill, director of the Cape Observatory in South Africa: "I was able to show Lord Rosse more stars in the Orion Nebula in a quarter of an hour than he ever saw in his life before with the 3-foot or 6-foot." The dominance of the Leviathan was ended, and telescopes of more modern type were set to take over.

There was also another episode which damaged the reputation of reflectors very badly. No major telescope had so far been sited in the southern hemisphere, and in Australia plans were made to install a 48-inch reflector at the Melbourne Observatory. By then Lassell's 24-inch had been succeeded by his 48-inch which, unlike the Leviathan, had access to the whole of the sky, and so it seemed logical to construct something similar for Melbourne; a 48-inch mirror was practicable, whereas a 48-inch objective was not (and this is still true today). A Royal Society committee was set up to give advice, and came to a decision which ultimately proved to be disastrous. Instead of favouring one of the new silver-on-glass mirrors, they opted for a mirror made of speculum metal.

One can follow their reasoning. The metal mirrors made by Rosse, Lassell and others had been tested and found to be good, whereas there had been no experience with glass mirrors of comparable size. Against this, it is a relatively easy matter to re-silver a glass mirror, but a metal mirror has to be heavily re-polished when it becomes tarnished, which is much more difficult and involves possible damage to the figure. Of the Royal Society committee members, only Lord Rosse tended to favour glass. A second mistake was the decision to do away with a dome, and house the telescope in a shed with a run-off roof.

The Grubb firm duly made the mirror, and also the mounting; in 1868 the telescope was inspected by the Committee, and was said to be excellent both optically and mechanically. Lassell's report was glowing: "The entire instrument is a great triumph of mechanical engineering and optical skill, and with the advantages of efficient working and a fine atmosphere, I trust it will add something to our knowledge of the heavenly bodies".

By August 1869 the telescope was in use at Melbourne, but it was not the success that had been expected. There were problems with the optics, and at one stage it was said that a star image looked rather like an ace of clubs. Neither was the run-off roof arrangement satisfactory, and there was constant trouble from wind-shake. Some visual work on nebulae was undertaken, and a little photography, but not much else, and the telescope was used only spasmodically. In 1904 an acid comment was made by G.W. Ritchey, by then probably the best mirror-maker in the world:

> I consider the failure of the Melbourne instrument to have been one of the greatest calamities in the history of instrumental astronomy; for by destroying confidence in the usefulness of great reflecting telescopes, it has hindered the development of this type of instrument for nearly a third of a century.

In fact the telescope was eventually put to good use. When the Melbourne Observatory was closed, in

An old engraving of the great reflector at Melbourne, which was one of the last large telescopes to have a metal mirror. For all sorts of reasons, this telescope turned out to be a disastrous failure

1945, the 48-inch was removed to the Mount Stromlo Observatory at Canberra. It was housed in a proper dome and given a new 50-inch glass mirror, since when it has been in constant use – but one has to admit that not much of the original Great Melbourne Reflector is left, so that Ritchey's condemnation was largely justified.

The Vienna telescope was followed by others. The Grubb firm was by no means alone; in America, Alvan Clark and Co. came very much to the fore, while in France there were the Henry brothers, Paul and Prosper. Another noted lens-maker was James Cooke, whose objectives are as fine as any ever made. During the 1880s and 1890s the great refractors were dominant; they included a 36-inch at the Lick Observatory on Mount Hamilton in California, a 33-inch at Meudon in the Paris area, a 31-inch at Potsdam in Germany, and so on (*a list is given in Appendix 2*). Largest of all was the 40-inch at the Yerkes Observatory in Wisconsin, master-minded by George Ellery Hale, of whom more anon. The 40-inch is still the main telescope at Yerkes, and is in use on every clear night.

The 40-inch refractor at Yerkes

The Age of the Great Refractors

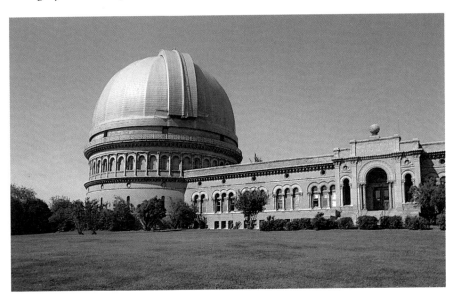

The Yerkes Observatory

A new factor had come into the reckoning. To an astronomer, "seeing" is all-important; the steadier and clearer the air, the better the view. High altitude is obviously an advantage, because there is less atmosphere above to cause trouble. Yerkes, true, is on the shores of Williams Bay, but this was because Charles Yerkes, the millionaire who financed it, insisted upon having the telescope in the general area of his home city, Chicago; but Mount Hamilton is well over 3000 feet high, and the Lick 36-inch benefited accordingly. Virtually all modern observatories are sited on mountain-tops, which may be inconvenient but is to all intents and purposes essential.

Seeing considerations were also paramount when Percival Lowell, a wealthy American amateur, made up his mind to establish a major observatory mainly to study the planet Mars, which he believed to be inhabited. Lowell and his assistants carried out an exhaustive survey of possible locations in the United States, and finally settled upon Flagstaff in Arizona, where the altitude is over 7200 feet and the observatory is well above the worst of atmospheric haze and dust. Lowell ordered a fine 24-inch refractor from Alvan Clark, and put it to good use even if he did not succeed in establishing the existence of the canals which he believed he had seen on Mars. I admit to having a special feeling for the Lowell telescope, because during my Moon-mapping days I was able to make extensive use of it – though by now, of course, the Observatory has acquired much

Lowell's 24-inch refractor

larger and more powerful reflectors, and the 24-inch is used mainly for public viewing.

There is something very attractive about a giant refractor. It may seem slightly old-fashioned, but to someone such as myself, interested mainly in the Solar System, it is unequalled. Yet with very few exceptions, all the world's largest refractors are old. The Innes 26-inch at Johannesburg, the last telescope made by the old Grubb firm, was brought into use in 1926, and has been noted for its work on double stars, but its recent career has been rather chequered, and at one stage it was put out of action altogether (I think I can claim to have had a hand in its restoration). But the Innes telescope is in Johannesburg city, where light pollution is a serious hazard; the sky is never really dark, and this means that the main advantage of a large instrument – the ability to detect very faint objects – is lost.

There was another reason why refractors began to give way to reflectors by the turn of the century. There

The Age of the Great Refractors

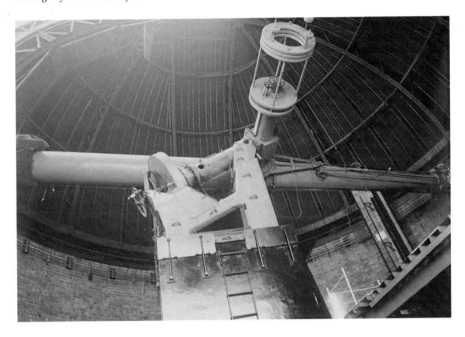

The Innes telescope at Johannesburg

The Great Paris Telescope – a 49-inch reflector – at the 1901 Paris Exposition

is a limit to the size of a working objective, because a lens has to be supported all round its edge, and if it is too heavy it will sag and distort under its own weight, making it useless. We must also concede that although the false colour nuisance can be greatly reduced, it can never be completely cured.

The largest refractor ever made had a 49-inch lens. It was built for display at the Paris Exposition of 1901, but the 180-foot tube could not be mounted in the

conventional way; instead it was left horizontal, and the light was brought to it by way of movable mirrors. Not surprisingly, it never worked well, and it was never used for any form of research; before long it was dismantled, and it is said that the object-glass remains stored somewhere in Paris. The fate of a Grubb 41-inch, destined for the Pulkovo Observatory in Russia, was even worse; before the optics could be installed, the mounting had rusted away.

By the end of the nineteenth century the whole situation was changing. The great refractors were in use, but it had become clear that the future of astronomical research lay with the reflector.

Chapter 8
George Ellery Hale and the Hooker Telescope

ASTRONOMICAL photography began in the 1840s, and within a few decades had become an important branch of research. The first good photographic atlas of the Moon was produced in Paris near the end of the century, and by then there had also been very good images of objects such as star-clusters and nebulae. Telescope drives had to be very good if time-exposures were to be made; electric drives lay in the future, but clockwork was adequate, and so were drives of the falling weight variety. Measuring instruments such as micrometers, attached to telescopes, also needed reliable drives, and of course spectroscopy was coming very much to the fore.

The Sun posed problems of its own. There was plenty of light to spare – in fact, almost too much; to look directly at the Sun through any optical device is remarkably dangerous, and will probably result in permanent blindness. Sunspots can be followed by the method of projection, as Galileo had shown long before, but for useful solar research it was essential to use instruments based on the principle of the spectroscope.

The pioneer studies of the solar spectrum were carried out as early as 1814, by Fraunhofer. He found that when a spectrum was produced, by passing sunlight through a prism, the result was a rainbow band, from red at one end through to violet at the other crossed by dark lines which were permanent in both position and intensity. It was later established that each "Fraunhofer line" is the trademark of one particular substance or group of substances, so that it became possible to find out which elements were present in the Sun. In 1891 a young American, George Ellery Hale, invented a new type of instrument: the spectroheliograph, which could produce images of the Sun in the light of one selected wavelength only, so that, for instance, a complete picture of the hydrogen or calcium distribution could be built up.

In 1892 Hale was appointed Professor of Astronomy at the University of Chicago, and began to plan the establishment of a new observatory. Money was needed, and Hale enlisted the aid of a local millionaire, Charles T. Yerkes, who provided the finance for a 40-inch refractor which, as we have already noted, was (and remains) the largest refractor in the world. The Clark objective was of superb quality, but Hale saw that so far as refractors were concerned the limit in size had been reached.

In his early career Hale was primarily a solar observer; he was the discoverer of magnetic fields in

This solar tower is on Mount Wilson. It is 150 feet tall

The dome housing the 60-inch reflector on Mount Wilson. The 60-inch itself is shown on the front cover

sunspots, and he made a number of important theoretical advances. He also designed solar telescopes, of which the first, financed by a Miss Helen Snow, is still operating at Mount Wilson in California. But to look into the depths of the universe Hale needed a much larger telescope, and it had to be a reflector. His constant call was for "More light!".

He aimed for a 60-inch. Again he was lucky; he approached the Carnegie Foundation, a powerful financial trust, and was promised enough money to build the 60-inch. George Willis Ritchey, at that time unrivalled as a mirror-maker, was put in charge of the optical side of the venture, and work began in earnest.

The choice of a site was of vital importance. Altitude would be an advantage, and there were two mountains in California which looked as though they might be suitable. One was Mount Wilson, overlooking the city of Los Angeles; the other was Palomar, much further away from any major centre of population. Palomar was certainly the better, but it was much more difficult to reach (there were few cars in those days), and so Mount Wilson was chosen. The 60-inch was finished in 1908. It was not so large as Lord Rosse's Leviathan, but it was of completely different type; it had a silvered glass mirror, it could reach any part of the sky, its mounting was amazingly precise, and it had a very accurate drive. The mounting was a skeleton construction (*see the front cover of this book*). After all, the only use of a telescope tube, whatever its design, is to hold the optics in the correct positions.

Even before the 60-inch came into operation, Hale was making new plans. Why not a 100-inch reflector? Once again he cast around for a financial backer, and once more he found one: John D. Hooker, a Los Angeles millionaire. Making such a colossal telescope was a daunting task, and when the mirror blank had been cast it took Ritchey and his team six years to figure it to the required optical curve. The mounting was of the so-called English type; the telescope was mounted on a massive beam fixed at both ends, and it was extremely stable, the main disadvantage being that the telescope could not be turned to regions close to the celestial pole.

The 100-inch Hooker reflector was the largest telescope in the world for thirty years

One evening in November 1917, Hale and three companions turned the 100-inch Hooker reflector to the sky for the first time. They focused on the planet Jupiter, and were appalled at what they saw – a shimmering, blurred image, lacking any sort of detail. For a moment it seemed that the telescope was doomed to be a failure, but the answer was straightforward enough; the dome had been left open in the afternoon, so that the mirror had become warm, and had not had enough time to cool down and regain its correct shape. A distortion of a tiny fraction of a millimetre would be too much. Hours later they tried again, looking at the star Vega – to find that the telescope was as perfect as even Ritchey could have hoped.

For the next thirty years the 100-inch was not only the largest telescope in the world, but was in a class of its own, so that it could show objects much too faint to be seen with any other instrument. It was with the 100-inch that one of the greatest of all astronomical discoveries was made: the realization that the starry nebulae, listed by Messier and drawn so painstakingly by Lord Rosse, are indeed independent galaxies.

Measuring star distances was never an easy matter. F.W. Bessel had had the first success, in 1838, when he had found that the dim star 61 Cygni is around 11 light-years away. (One light-year is the distance travelled by a ray of light in one year: approximately 5.8 million million miles.) Bessel had used the method of parallax, which in principle is exactly the same as the procedure followed by a surveyor who wants to measure the distance of some inaccessible object, such as a mountain-top. The object is viewed from the opposite ends of a baseline, and the slight difference in direction makes it easy to work out the required distance by simple mathematics. For 61 Cygni, Bessel's baseline was the diameter of the Earth's orbit: 186,000,000 miles. The star shifted in apparent position by less than 0.3 of a second of arc, but this was enough for Bessel to obtain a very accurate result.

Unfortunately the parallax method works out to only a few hundred light-years; further out, the parallax shifts become so small that they are swamped by unavoidable errors in observation – and the spirals were much further away than that. However, there was a solution, thanks to some convenient stars known as Cepheid variables. The name comes from Delta Cephei, the first-discovered member of the class, which lies in the far north of the sky.

Unlike normal stars, Cepheids do not shine steadily. They brighten and fade with clockwork regularity, and it had already been found that the real luminosity of a Cepheid is linked to its period of variation; the longer the period, the more luminous the star. Thus Delta Cephei itself, with a period of just over 5 days, is less powerful than Eta Aquilae, where the period is over 7 days. When the true luminosity is known, the distance can be found – once allowance has been made for various complications, such as the absorption of light in space.

Edwin Hubble, one of the astronomers at Mount Wilson, set out to locate Cepheids in the spirals. Only the 100-inch was powerful enough for the purpose, but in 1923 Hubble was able to identify Cepheids in several of the spiral systems, including Messier 31 in Andromeda. As soon as he measured their periods, he realized that they were much too remote to be members of our Galaxy. His derived distance for the Andromeda Spiral was 900,000 light-years, later reduced to 750,000 light-years. In fact this was later found to be an under-estimate, but the main problem had been definitely solved.

The next advance was also made possible by the light-grasp of the 100-inch. Together with his colleague Milton Humason (who began his career at Mount Wilson as a mule driver, and ended it as one of the world's leading astronomers), he studied the spectra of the objects which he now knew to be galaxies. The spectrum of a galaxy is made up of the combined spectra of many millions of stars, and is bound to be something of a jumble, but the main Fraunhofer lines can be made out, and their positions measured. According to the well-known Doppler effect, the apparent wavelength of a light-source is affected by its motion relative to the observer; with an approaching object all the spectral lines will be shifted over to the blue or short-wave end of the rainbow band, while with a receding object the shift will be to the red. Hubble and Humason found that apart from a few of the very closest systems, including the Andromeda Spiral, all the shifts were to the red, so that every group of galaxies was racing away from every other group. In fact this had already been noted by V.M. Slipher, using the Lowell 24-inch refractor, but its significance had not been appreciated.

This is not all. The further away a galaxy is, the faster it is receding. There is a definite law, known as the Hubble Constant, so that once the red shift has been found the distance automatically follows. Today we can identify systems so far away that their light takes well over 10,000 million years to reach us.

Doubts remain, and the precise value of the Hubble Constant is still uncertain; there are even suggestions that the shifts may not be due entirely to the Doppler effect, in which case there will be a need for dramatic re-thinking. But whether or not this turns out to be the case, the basic principles go back to the pioneering work carried out by Hubble and Humason with the 100-inch Hooker reflector.

Of course, all this work was carried out photographically. More than a century ago the human eye had been superseded by the sensitive plate; today photography is itself being supplanted by electronic devices, and these can be used with the 100-inch, which is therefore even more efficient now than it was when first built. It has also been fully computerized. Against this, however, is the growing problem of light pollution. Mount Wilson is inconveniently near Los Angeles, and for a while, in the 1980s, the situation became so bad that the telescope was temporarily mothballed. It has now been brought back into full use, and although the skies are less dark than they used to be an extensive research programme can still be carried out. The 100-inch is no longer the largest of all telescopes, and indeed is not even in the top twenty, but it has a special place in the history of science. It, and it alone, could be used to show that our Galaxy is but one of many star-systems spread through the vastness of space.

Chapter 9
Palomar

THERE have been a few telescopes which have been responsible for particular advances in our knowledge. Galileo's refractor was the first; then Newton's original reflector; then the Rosse Leviathan, and in the twentieth century the 100-inch at Mount Wilson. Next in sequence comes the Palomar 200-inch, which again was the brainchild of that remarkable man George Ellery Hale.

Hale was never satisfied; even after the triumph of the 100-inch he yearned for something even larger – a 200-inch this time. As usual, finance was the first hurdle, and once more Hale overcame it. He went to the Carnegie Institute of Washington, and obtained an initial grant of 6,000,000 dollars, with the promise of more to come. Two immediate decisions had to be made. First, where should the new telescope be sited? And secondly, of what material should the main speculum be made?

Years before, Hale had considered Palomar Mountain, and had rejected it only because of its inaccessibility. Since then transport had become much easier, and Palomar was certainly a much better site than Mount Wilson: so Palomar was chosen. It is over 7000 feet high, and well away from any serious light pollution (though the situation today is not as perfect as it used to be). So far as the mirror was concerned, quartz would have been ideal, but after a long series of experiments its was reluctantly decided that the problems of making a quartz mirror were too great, and pyrex was used instead.

The first attempt at casting the blank was made on 25 March 1934. The tank containing 65 tons of molten

The author, standing in front of the 200-inch reflector

pyrex took more than two weeks to fill, and a further sixteen days to heat to the required temperature, after which there was slow, regular cooling to room heat. When the blank was ready, it was found that there were internal strains in the glass. Wisely, it was decided to keep the first blank as a spare, and try again. This time all went well, and the blank was taken to Pasadena in California to be figured. It was of course to be used as a Cassegrain, and Hale commented that the hole in the centre of the speculum was as large as the entire objective of the Yerkes refractor.

Figuring the mirror proved to be even more difficult than had been anticipated, and it took a very long time. Hale, sadly, did not live to see the 200-inch completed; he died in 1938, and there were further delays when the United States was drawn into the war in 1941. So it was not until 3 June 1948 that the 200-inch, appropriately named in honour of Hale, was dedicated. At the inaugural ceremony Dr. Lee DuBridge, President of the Carnegie Institute, made some apt comments: "This great telescope before us today marks the culmination of over two hundred years of astronomical research. For generations to come, it will be a key instrument in Man's search for knowledge."

Drive up the winding road to the top of Palomar Mountain, and you will come within sight of the huge dome with surprising suddenness. It is 137 feet in diameter, and weighs 1000 tons. Go inside, and the first thing you will see is what looks like a giant horseshoe. This is the mounting of the telescope – very different from the design adopted for the 100-inch and although it weighs 500 tons it can be moved around slowly and with absolute precision. Inside it is mounted the telescope itself, which, like the 60-inch and the 100-inch, has a skeleton "tube", with the main mirror at the bottom. The tube is so large that when using the prime focus system there is no need for a secondary mirror; the observer can sit in a cage actually inside the telescope. The main speculum is coated with a very thin layer of aluminium so as to make it as reflective as possible; every scrap of light has to be collected.

Before long it was clear that the Hale telescope would be just as dominant as the Hooker reflector had been in its time; it too was in a class of its own, and the

The dome of the 200-inch Hale reflector

The Hale 200-inch reflector. The massive horseshoe mounting weighs around 500 tons

first major discovery with it was made in 1952. Hubble was coming to the end of his career – he died in 1953 – and his mantle fell upon Walter Baade, a German astronomer who had emigrated to America before the war and had joined the Mount Wilson staff as early as 1931. Like Hubble, he concentrated upon studies of galaxies, and the 200-inch increased his scope enormously. In 1952 he discovered that there has been a serious error in the Cepheid scale. There are two different classes of short-period variables, and they have different period-luminosity laws; Type I Cepheids are much more powerful than those of Type II, and Hubble, through no fault of his own, had used the wrong type to make his estimates of distance. The variables which he had been using as "standard candles" were twice as luminous as he had believed, so that they also had to be twice as far away. In one short research paper Baade calmly doubled the size of the universe. The distance of the Andromeda spiral was not a mere 750,000 light-years; the real value is well over 2,000,000 light-years.

The extra light-grasp of the 200-inch meant that Cepheids could be detected in galaxies which were much further away than those of the Local Group; the telescope could also show details quite beyond the range of any other instrument. Moreover, there was another factor also to be taken into consideration: improvements in ground techniques.

Electronics began to take over. What is termed a CCD or Charge-Coupled Device is much more sensitive than any photographic plate, and the 200-inch was well able to adapt to these new techniques. By now the observer no longer has to spend long hours in a cold, lonely observing cage; he can sit in the comfort of a control room, a cup of coffee by his side, and watch the results coming through on a television screen. I well remember a talk I once had with Edwin Hubble himself. He told me that on one occasion he had spent two whole nights with the 100-inch at Mount Wilson, obtaining the spectrum of a single galaxy. Now, the same result can be obtained in a couple of minutes. Also, computerization has taken over, and this too was no more than a vague dream when the Hale reflector first began its career.

The 200-inch, like the 100-inch, was designed for "deep space" research, and was not concerned with the Solar System. A few photographs of planets were taken, mainly to satisfy public curiosity, but that was all. There were much more important things for the telescope to do.

Of course, the 200-inch is by no means the only major instrument at Palomar; for example, there is a fine 60-inch. There is also a Schmidt camera, which is of the greatest importance.

All very large telescopes have one common weakness: their fields of view are very small, so that to make a photographic atlas of the whole sky would take a very long time indeed. In 1930 an entirely new optical system was developed by an Estonian researcher, Bernhard Schmidt (he was originally an explosives expert, but blew part of his arm off during an experiment, and decided to turn to optics as being rather safer). On the Schmidt system, the incoming light is collected by a mirror which is spherical, not parabolic

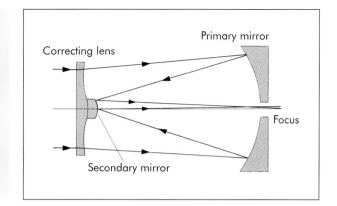

Principle of a Schmidt-Cassegrain telescope

as with an ordinary telescope. Normally this would introduce serious distortions, but these are removed by means of a thin glass plate, with a very complex profile, fixed at the upper end of the tube. The photographic plate is situated at the prime focus. The main problem is that the focal plane is curved, and so a special plateholder has to be used to bend the plate into the correct form. The end result is that sharp star images can be obtained over very wide fields, often well over 10 degrees across. The Oschin Schmidt camera at Palomar has a 48-inch mirror; it was completed at about the same time as the 200-inch, and has been in constant use ever since. Only the 53-inch Schmidt at the Karl Schwarzschild Observatory at Tautenberg, in Germany, is larger.

By now the 200-inch is regarded as an old telescope, and certainly it is senior in age to any other telescope of comparable size, but it is none the worse for that, and it is worth noting here two more investigations in which it has played a leading rôle.

Light, as we have seen, is a wave motion. Wavelengths longer than that of red light do not affect our eyes; we have infra-red, then microwaves and then radio waves. Radio astronomy goes back to the early 1930s, when Karl Jansky, a Czech-born radio engineer living in America, was using an improvised aerial to study "static" when he found that he was picking up radio emissions from the Milky Way. Curiously, he never really followed up his discovery as he might have been expected to do, and radio astronomy did not attract a great deal of attention until after the war, when the first intentional radio telescopes of large size were built. Perhaps the term "radio telescope" is misleading; an instrument of this kind is really in the

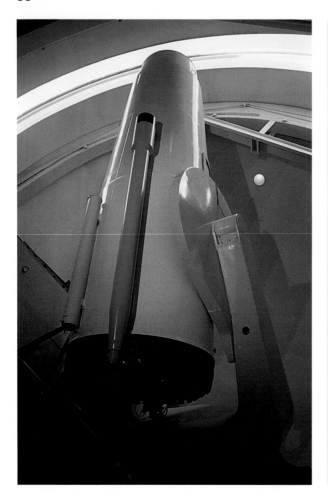

The 48-inch Oschin Schmidt telescope, at Palomar

nature of a large aerial, and one certainly cannot look through it. The usual end product is a trace on a graph, which may not look exciting, but which can provide us with information which we could never obtain in any other way. Some radio telescopes are in the form of "dishes", of which the most celebrated example is the 250-foot Lovell telescope at Jodrell Bank in Cheshire; others look more like collections of barbers' poles, but all have essentially the same purpose.

An initial surprise was that the radio sources in the sky are not, in general, associated with objects which are prominent visually (with certain obvious exceptions, such as the Sun). It was found that the strongest sources came from areas which seemed at first inspection to be more or less blank. It emerged that some kinds of galaxies are unusually powerful in the radio range, and there were also supernova remnants such as the Crab Nebula in Taurus, known to be the remains of a star which had flared up in the year 1054. (The Crab is 6000 light-years away, so that the actual outburst occurred in prehistoric times.) A catalogue of radio sources was compiled at Cambridge, with the aid of the radio telescopes there, and in 1962 radio astronomers at Parkes, in Australia, were able to identify one particular source with what seemed to be a faint bluish star. The results were sent to Palomar, where Maarten Schmidt (no relation to Bernhard!) used the 200-inch to examine the spectrum of the object. To his surprise, he found that it was not a star at all, but something much more dramatic. The spectrum showed lines due to hydrogen, so greatly red-shifted that the object had to be very remote indeed. This was the first quasar, a term derived from the cumbersome "quasi-stellar radio source". Quasars are indeed remote; they are far more powerful than normal galaxies, so that they can be seen over immense distances. It now seems that a quasar is the core of a very active galaxy, and that the

The Lovell radio telescope at Jodrell Bank

power source comes from a gigantic black hole at the centre of the system. Thousands of quasars are now known; by no means all of them powerful in the radio range, but only the Palomar reflector was able to identify the very first.

Then, too, there was the return of Halley's Comet. This is the only bright comet which comes back to perihelion regularly; it has a period of 76 years, and has been seen at every return since that of BC 240. It was at perihelion in 1835 and again in 1910, so that it was due once more in 1986. Searches for it began well ahead of time, and developed into quite a race! Fittingly, the first sighting came from Palomar, where the comet was recorded with the 200-inch on 16 October 1982 as a very dim object, below magnitude 24. It was badly placed at the 1986 return, and will be no better when it comes back in 2051, but if you could see it at its next visit, in 2127, you would be treated to the spectacle of a brilliant comet with a long tail, casting strong shadows. Unfortunately, I fear that no reader of the present book will have that privilege.

The 200-inch is still in action on every clear night, and there is no reason why it should not continue to operate indefinitely. It introduced a new era in the story of telescopic astronomy, and perhaps its best work still lies ahead.

Chapter 10
The New Telescopes

IT was sometimes suggested that with the Palomar 200-inch the limit of size had been reached, and that the advantages of any larger telescope would be more than cancelled out by increased problems of atmospheric turbulence. Certainly the next giant instrument was not a success. In 1976 the Russians set up a 236-inch reflector at Mount Pastukhov, in the Caucasus region, and great things were expected of it.

Sadly, these hopes were not realized. There were difficulties with the mirror, which was so massive that it was seldom operating at the correct temperature to maintain the optical curve. Secondly, the site was poor. There are no suitable mountains in what was then the Soviet Union, and the climate in the Caucasus region is not good. Very few results have come from the telescope, and yet it was responsible for one major advance in technique: it was the first large modern telescope to be given an altazimuth mounting.

Before the development of computers, it was hopeless to mount a great telescope on an altazimuth, because there are two separate movements to be made – north/south, and east/west – and it would have been impossible to attain the necessary accuracy. But computers were developed during the 1960s, and by the time that the Russian giant was ready it had become practicable to provide it with a computerized system which would make guiding as easy as with a conventional equatorial. Moreover, the altazimuth has major advantages. The mirror does not have to be swung about to the same extent as with an equatorial, and

The 236-inch reflector at Mount Pastukhov has not been a great success, although it pioneered the use of altazimuth mountings for large telescopes

observing procedure is considerably simplified. With the Russian reflector, the altazimuth pattern proved to be highly satisfactory, so that even if the telescope under-performed optically it cannot be dismissed as a total failure. Since then, almost all large telescopes have been mounted on altazimuths.

This is not Russia's only large telescope; at the Crimean Astrophysical Observatory there is a good 102-inch reflector, and there is also a 104-inch at Byurakan in Armenia. In pre-war days the main Soviet observatory was at Pulkovo, outside what was then called Leningrad, where the main telescope was a large refractor; but during the siege of Leningrad the observatory was destroyed by the Germans, and when it was rebuilt no really major telescope was put there. As one Russian astronomer told me, there are definite problems. In summer the sky is too light for any real work to be done, and during winter the cloud cover is almost constant, so that observing is more or less confined to spring and autumn!

The New Telescopes

The dome of the 236-inch telescope on Mount Patsukhov

Until relatively recent times almost all the world's large telescopes had been set up in the northern hemisphere. Yet some of the most important objects in the sky lie in the far south, so that they can never be seen from the latitudes of Europe or most of the United States. For example, the Clouds of Magellan – the

The dome housing the 102-inch telescope at the Crimean Astrophysical Observatory

nearest major galaxies, less than 200,000 light-years away – never rise from anywhere north of about latitude +30 degrees, and neither can Europeans see the Southern Cross, the magnificent nebulae in the constellation of Carina, or the two brightest globular clusters, Omega Centauri and 47 Tucanae. It was realized that there was an urgent need for research telescopes in the southern hemisphere, and during the 1970s several were set up.

The first of these was the AAT or Anglo-Australian Telescope at Siding Spring, near Coonabarabran in New South Wales, well north of Sydney. The main headquarters of Australian astronomy had been at Mount Stromlo, outside Canberra, but the usual problems of light pollution had become unpleasantly obvious, and a new site was needed. There are no high peaks in Australia; the Warrumbungles of New South Wales rise to little more than 4000 feet, but seeing conditions there were tested and found to be at least adequate. There is no real trouble from artificial lights, and the Observatory is very accessible, though driving up the mountain road after dark can be a somewhat hazardous business; kangaroos have absolutely no road sense, and are always apt to bound out without warning.

The AAT has a 153-inch mirror (or 3.9 metres, if you prefer metric), weighing 16 tons; it is made of Cervit, which is very suitable because its size and shape barely alter with changes in temperature. The optical curve of the mirror is accurate to within a few millionths of a centimetre, and is coated with a layer of aluminum 1/10,000 of a millimetre thick. The focal length of the

Siding Spring in Australia, showing the observatory domes

The New Telescopes

The AAT has a 153-inch mirror made of a ceramic material with a very low coefficient of expansion

primary mirror is 42 feet – a far cry from the unwieldy aerial telescopes of long ago.

As with all modern telescopes, the AAT can be used according to one of several optical systems; prime focus, Cassegrain and Coudé, in which a series of five mirrors brings the incoming light to a fixed focal point. This means that heavy equipment can be used without having to be moved around, which is an important consideration. The mounting of the telescope is equatorial, and in fact the AAT was one of the very last major telescopes to be built on the old-style mount.

The AAT is completely automatic. The observer sits in the adjacent control room, watching a television screen; to set the telescope on to the required target, the co-ordinates are plugged in, and the telescope swings smoothly to the indicated position. The setting is accurate to within a couple of minutes of arc, and a slight manual adjustment is enough to bring the target into the centre of the field of view. Then the driving mechanism takes over, and work can continue without any danger of the target drifting out of the field.

Of course the very latest equipment is used, but, unusually for so large an instrument, old-fashioned photography is also undertaken, mainly by David Malin, who sits in a cage at the prime focus and has produced images which are unquestionably the very best ever obtained with an Earth-based telescope. Also at Siding Spring, in a dome only a few yards from that of the AAT, is the UKS or United Kingdom Schmidt, with a 49-inch mirror. This is the same size as that of the Oschin Schmidt at Palomar, and the two have combined to draw up an all-sky atlas; the UKS reaches southern regions forever denied to the Oschin, while the UKS is similarly unable to reach the area of the north celestial pole.

Insofar as published results are concerned, it seems that the AAT and the UKS are the most productive of all modern telescopes. Moreover, the accessibility of Siding Spring means that when any maintenance is needed, it can be carried out at once, without the need for travelling for a long way over inhospitable terrain. All in all, the Siding Spring Observatory may not be the world's best site, and it does not have the world's most powerful telescopes, but in efficiency and output it is second to none.

Southern-hemisphere astronomy really began in South Africa, and the first Cape observatory dates back to the 1820s. It was to the Cape that John Herschel went in the 1830s, taking the 20-foot reflector to carry out a

The United Kingdom Schmidt (UKS) at Siding Spring in Australia

The 74-inch Sutherland reflector in South Africa

detailed survey of the southern stars. The Cape Observatory is still active, but seeing conditions there are not good, and by now the main South African telescopes have been concentrated at Sutherland in Cape Province, where conditions are much better. A major decision about this was taken some years ago. There were already observatories at Pretoria, Johannesburg and Bloemfontein, and it took courage to relegate these to minor rôles; indeed, the Radcliffe Observatory outside Pretoria was closed down altogether – the main telescope, a 74-inch reflector, was transferred to Sutherland, and the Radcliffe site was simply abandoned. (The 27-inch Grubb refractor at Johannesburg was left *in situ*, because it did not fit in with the general observing programmes.) Plans are now being made for a really large South African telescope, but as yet these negotiations are still in a very early stage.

The cylindrical dome housing the Sutherland reflector

The South African decision underlined the importance of "seeing". Parts of the United States are good, and many large telescopes have been erected; the official U.S. observatory is at Kitt Peak in Arizona, where the main optical telescope is the 150-inch Mayall reflector. Here too is the McMath Solar Telescope, which looks rather strange at first sight. The sunlight is caught at the top of the instrument by a rotatable

Sunset at Sutherland

The New Telescopes

The McMath solar telescope at Kitt Peak in Arizona, currently the most powerful of all solar telescopes

mirror or coelostat, and is then sent down the long, inclined tube to a second mirror at the bottom; this reflects the light back to yet another mirror at the "half-way" stage, after which the light is directed into the instruments in the laboratory. The McMath is the most powerful of all solar telescopes, and is so versatile that it can also be used for stellar work. The tube is inclined solely for convenience; most solar telescopes, such as those at Mount Wilson, are towers.

Of all the southern hemisphere sites, probably the best are to be found in Chile. The Atacama Desert is one of the most inhospitable places on earth, but the air is steady, the altitude is high, and the percentage of clear nights is very great. In the general area of the town of La Serena – though sufficiently distant to avoid any trouble from light pollution – there are three major observatories: La Silla, Cerro Tololo and Las Campanas. The first of these is run by the European Southern Observatory, with its headquarters in Germany, and the other two by the Americans.

La Silla is remarkable; on the mountain-top, domes seem to sprout like mushrooms. There are fifteen in all, plus a 49-foot radio telescope of the dish variety, and the Observatory grounds cover a large area amounting to 250,000 square miles, which, believe it or not, was purchased in the 1960s for the princely sum of 10,000 dollars. The largest telescope is the conventional 141-inch; there are also the 87-inch Max Planck reflector, the 60-inch Danish telescope, a 39-inch Schmidt and various telescopes of smaller aperture. A sky survey carried out by the La Silla Schmidt and the UKS in Australia has resulted in the detection of 15,000 new galaxies as well as many star-clusters, asteroids, and a number of comets.

The most interesting telescope at La Silla is the NTT or New Technology Telescope, with a mirror 138 inches across. It is short and squat, with a very short focal length indeed; it is of course mounted on an altazimuth. When it is rotated, the dome rotates with it, so that the slit is always in the right position.

The heart of a telescope is its main mirror, and that of the NTT is very thin, with a maximum thickness of only 24 centimetres; it is made of the low-expansion glass ceramic Zerodur, and the whole telescope floats on a layer of oil 0.3 of a millimetre thick. Swinging a large mirror means distorting it, and this would be disastrous with a mirror as wafer-thin as that of the NTT; therefore, two new techniques are used to keep the figure perfect. The first is termed active optics. In this, the shape of the mirror is continually modified by computer-controlled pads behind it; with the NTT there are no fewer than 78 pads, and compensation can be carried out in a tiny fraction of a second. With adaptive

The New Technology Telescope at La Silla. The mirror is 138 inches in diameter

The author standing in foreground of the NTT dome

optics, a small computer-controlled mirror is inserted in the optical system. By monitoring a relatively bright star in the field of view, the shape of this small mirror can be continuously modified, removing distortions due to turbulence in the Earth's air. At the time of writing (January 1997) the NTT is temporarily out of action while the very latest equipment is being installed.

Another development is that the NTT can be operated by remote control. The control room at Garching in Germany, headquarters of the European Southern Observatory, is a twin of the control room in the NTT building, so that when an observer wants to use the telescope he has no need to make the long and expensive journey to Chile; he can operate the telescope from his own office. Gone are the days when the observer had to sit at the telescope and guide it. Nowadays, he need not be in the same building – or even on the same continent. Times have changed.

Cerro Tololo, with its 158-inch reflector, is more "condensed" than La Silla, and all the main domes are set up on a smallish plateau. Thirdly there is Las

Cerro Tololo in Chile showing two of the observatory domes

Campanas, where the main telescope is the 100-inch Irénée du Pont reflector. In the near future there will be yet another observatory, this time at Cerro Pachón in the same general region.

At the moment the world's southernmost observatory is at Mount John on the South Island of New Zealand, but within the next few years a major observa-

The dome housing the Mount John reflector in New Zealand

At present, the world's southernmost observatory at Mount John in New Zealand, where this 24-inch reflector is used

tory is to be set up actually at the South Pole. The fact that the day and night are each six months long will have definite advantages, and the seeing conditions are known to be excellent. The main drawback is that the climate is a little chilly – in fact, it is too cold even for penguins!

Chapter 11
The Summit of Mauna Kea

There is no doubt that the Earth's atmosphere is the main enemy of the observational astronomer. The only solution is to take the scientific equipment as high as possible, which is why mountain-tops are so favoured. But without involving space research methods, how high can one go?

Think of Hawaii, and you conjure up a picture of sun-soaked beaches and bikinis. True, parts of Hawaii are like that, but others are not, and in 1963 Gerard Kuiper, a Dutch astronomer who spent most of his career in the United States and was a pioneer in lunar and planetary observation, cast his eye upon Mauna Kea, on the Big Island. Mauna Kea ("White Mountain") rises 32,000 feet

Domes at the Mauna Kea observatory in Hawaii. The site is 14,000 feet above sea level and seeing conditions are excellent

The Summit of Mauna Kea

from the ocean floor to an altitude of almost 14,000 feet above sea-level; it is extinct, though its neighbour, Mauna Loa, it highly active. On top of Mauna Kea the air is very dry. It is also very thin, so that one's lungs take in only 39 per cent of the normal amount of oxygen, and violent exercise is not to be recommended; there are some people who cannot tolerate the altitude at all. (I am lucky, mainly because I was a wartime flyer; altitude does not affect me.) Kuiper realized that in spite of its obvious drawbacks, Mauna Kea would be an excellent site for an observatory. He met with considerable opposition, but in the end his arguments prevailed, and today Mauna Kea is the site of a whole collection of telescopes, including the largest so far in operation.

Driving up to the summit from Hilo, the largest town on Big Island, is quite an experience. Saddle Road runs between the two giant volcanoes, Mauna Kea to one side and Mauna Loa to the other, and it is a long, steady climb, though the gradient is not particularly steep. At just below 10,000 feet you come to Hale Pohaku ("Stone House") where astronomers using the telescope stay during their observing runs; nobody actually sleeps at the summit, because the thinness of the air makes it dangerous. Driving from Hale Pohaku to the Observatory takes a mere twenty minutes or so, and nowadays the road is tolerable, though four-wheel drives are very much to be recommended. Suddenly you come within sight of the first of the domes, and the view is breathtaking.

Britain is well represented by the presence of UKIRT, the United Kingdom Infra-Red Telescope. Infra-red radiations from space are absorbed by water vapour in the air, and there is not much atmospheric water vapour at a height of 14,000 feet, so that conditions are as good as they can be from the Earth's surface. Because infra-red waves are longer than those of visible light, a telescope designed to study them need not be as accurate as an optical telescope, and UKIRT's 150-inch mirror is very thin; in the event it proved to be so good that it could also be used for optical observations, but this was sheer bonus. It began operations in 1979. The mirror weighs a mere $6\frac{1}{2}$ tons, and the entire telescope mounting is decidedly lighter than with conventional instruments of the same size; as usual there are several optical systems, including a Cassegrain and a Coudé. The UKIRT is very versatile, but is particularly suited to studies of highly energetic galaxies thousands of millions of light-years away.

Also on Mauna Kea there is the IRTF, the NASA infra-red telescope (the letters stand for "Infra-Red Telescope Facility"); this has a 118-inch mirror. At a slightly lower level down the volcano there is the JCMT or James Clerk Maxwell Telescope, which is more in the nature of a radio telescope and is designed to pick up microwave radiations, which come in between the infra-red and the radio range.

The first telescope on Mauna Kea dates back to 1970, but many have been added since, notably the 141-inch Canada–France–Hawaii or CFH reflector. Others are planned. The giant Japanese Suburu reflector, with a mirror no less than 327 inches across, is expected to be ready within the next couple of years, and its dome is already in place. But all these pale beside the mighty Keck telescopes, which are already working.

Obviously there is a practical limit of the size of a single mirror, whether it is made conventionally or on Roger Angel's new spin-casting technique. Therefore when a new, very large telescope was planned for Mauna Kea, a new pattern had to be found. The mirror was to be 10 metres across, giving a clear aperture of 387 inches, and at the instigation of Dr. Jerry Nelson, of the University of California, it was decided to attempt a segmented mirror, made up of separate pieces of glass fitted together to make the correct optical curve. After all, as was pointed out, we have been making mosaic

Dome of the IRTF telescope at Mauna Kea

The Summit of Mauna Kea

The dome built to house the 327-inch Suburu reflector on Mauna Kea

floor tiles for many centuries, and the idea of a mosaic is not novel – except to astronomers.

Nelson first made his suggestion in 1977, and the segmented pattern was approved three years later, though not without considerable opposition. Money had to be found, and this was where the W.M. Keck Foundation came into the picture. It is one of the largest of those American organizations giving grants to charities and other good causes, and it donated £70,000,000, which amounted to 75 per cent of the total cost of the proposed telescope. The rest was found, and construction began.

The mirror is almost twice the size of that of the Hale reflector. There are 36 hexagonal segments, aligned by a control system which has to be accurate to within a millionth of an inch – a thousand times thinner than a human hair. The telescope structure is 81 feet high, and together with the mirror weighs almost 300 tons. The overall impression of the telescope is somewhat futuristic, and there are numerous wires, ladders and platforms, so that nothing could be less like the old-time reflectors, or the great refractors. It is hardly necessary to add that the mounting is of the altazimuth type.

The total weight of the glass in the mirror is only 14.4 tons, as against 41 tons for the single mirror of the

The Keck segmented-mirror telescope. Several of the hexagonal mirror sections are clearly visible

Russian 236-inch. Because the mirror had to be made in sections, normal testing could not be used until all the segments had been suitably bent; this was done by using stress techniques invented by Jerry Nelson specially for the purpose. Each segment is 6 feet in diameter and 3 inches thick, weighing 880 pounds; the material is Zerodur. The focal length of the whole telescope is 57.4 feet.

The control system, so vital to the success of the telescope, depends upon disk pads and what are called *whiffletrees* – a whiffletree being a vertical rod which branches out into twelve short posts (the name comes from a last-century pivoting crosspiece which allowed draft animals in a team to move independently while still pulling a cart). The whiffletrees and flex disks keep the mirror segments rigidly mounted, while still allowing them to be moved for aligning the mirror array.

Since the 36-segment array forms the correct optical curve, each individual segment has to have a special curvature which is not symmetrical; one can well imagine the amount of calculation involved. Note also that the mirror cell must hold its shape to within 1/25 of an inch, no matter where the telescope is pointing in the sky. Then, of course, there is the dome, 101 feet high and 122 feet wide, with a total weight of 700 tons.

"First light" came on 24 November 1990, when only nine of the segments were in place; even so, the light-grasp was already equal to that of the Palomar 200-inch. And when all the segments had been fitted, it was found that the telescope worked just as well as had been hoped. The decision to make a segmented mirror had been fully justified.

There was more good news to come. A second grant from the Keck Foundation made it possible to add a twin telescope, set up in a dome close beside the first. Quite apart from the fact that this doubled the available telescope time – which is always in great demand from would-be observers – it will eventually be possible to use the two Kecks together, as a single unit. In theory it would then be capable of distinguishing the two headlights of a car, separately, over a distance of 16,000 miles.

If a mirror the size of the Kecks can be made on the segmented pattern, can one go even further? There seems no valid reason why not, though of course the practical difficulties become greater and greater with each increase in size. At least the two Kecks have shown the way.

Chapter 12
From Sussex to La Palma

IT has often been said that Britons are paranoid about the weather, and to a certain extent this may be true; certainly astronomers are not enamoured of our climate, and all in all it is surprising that so much good observational work has been done from here. Yet only once, and then for a brief period, have we had a really large telescope. This was the Isaac Newton reflector, known popularly as the INT.

As we have noted, the Royal Observatory at Greenwich was founded as a navigational and timekeeping

Herstmonceux Castle

establishment, and it was only after the Neptune episode that the first reasonably large telescopes were erected there. The Thompson 26-inch refractor was brought into use in 1897, and was followed by various reflectors, including the Yapp 36-inch. Navigation and timekeeping still played major rôles, but the Royal Observatory became a true astronomical centre, and a great deal of observational and theoretical research was carried out.

Yet even in the early part of the twentieth century it was becoming clear that the future was bleak. London was growing, and becoming both dirtier and brighter; Greenwich Park was no longer the remote suburb that it had been in Newton's time. So the decision was made to shift the working telescopes to a new site at Herstmonceux in Sussex, not far from the little town of Hailsham, and to convert the Old Royal Observatory into a museum. The move was delayed because of the war, and took a long time, but by the mid-1950s it was complete, and the Royal Greenwich Observatory, Herstmonceux, was in full operation.

Obviously it would be advantageous to have a really large telescope. Quite apart from the publicity value (which is not to be ignored; after all, in the long run it is the taxpayer who foots the bill), a major instrument would be a great help to British observers. "Telescope time" is always at a premium, and there were times when British astronomers were forced to curtail their research programmes because no adequate telescope was available to them. A 96-inch mirror was obtained from America, and in 1967 the Isaac Newton Telescope was set up in the grounds of Herstmonceux Castle.

Without pretending that the telescope was as perfect optically as, say, the AAT, it was satisfactory enough, but the site was not. It was probably a mistake to erect the telescope in Sussex in the first place, though the decision was more political than purely scientific. The situation deteriorated still further with the growth of the nearest large town, Eastbourne, and finally the INT was moved to a new site on top of Los Muchachos, an extinct volcano in the Canary Islands. Here the altitude is over 7000 feet, and conditions are very good indeed. The other telescopes remained at Herstmonceux, apart from a 28-inch refractor, which was lifted bodily out of its dome and taken back to its original home at Greenwich.

However, it is not correct to say that the INT was shifted to La Palma intact. Far from it. To start with, it

The Equatorial Group of domes at Herstmonceux. The telescopes are still inside the domes, and will be brought back into full use

was given a new mirror, this time with an aperture of 101 inches. The mounting had to be completely altered because of the latitude difference between Sussex and the Canaries, and there had also to be a new dome, so that all in all not a great deal of the original INT was left. This gave me an idea, which I put forward officially to the Director at Herstmonceux, Professor Alexander Boksenberg (chief architect of the CCD). Herstmonceux retained a usable mirror, most of a mounting, and a suitable dome. Why not re-erect the telescope as cheaply as possible, and use it for as much research as could be managed, plus the testing of new equipment and possibly public display also?

The idea was well received, but there was the usual problem: money. With official blessing I contacted a Scottish engineering firm, and the managing director said that he was prepared to re-erect the telescope free of charge provided that it was given an appropriate name. A working committee was set up, and all seemed to be going well when, in 1986, the Science and Engineering Research Council (SERC), which controlled all the finances, decided to close Herstmonceux and transfer the Royal Observatory to a new site.

Predictably there was a storm of protest. Such a move would break up the efficient team at Herstmonceux; it would mean that the Library would be dispersed and the public exhibition closed; training of new observers would cease, with adverse effects on the very strong astronomy department at the University of Sussex; it would disrupt the programmes at La Palma, and it would cost money. I was asked to take part in the "resistance", and on 6 June 1986, again with official sanction, I called a meeting at the Scientific Societies Lecture Theatre so that the whole matter could be thrashed out. Of all those present – and the meeting was packed out – only the SERC representatives favoured the move, and among the main speakers against it were the Directors of the Cambridge and Greenwich Observatories. Meetings were also held in both Houses of Parliament, with support from both political parties. It was to no avail; the SERC laid down that the Observatory was to be transferred to what was really an office block in Cambridge. The telescopes

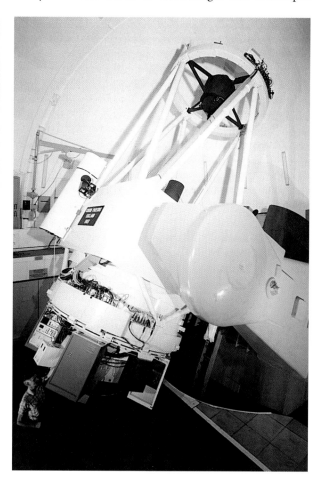

The Isaac Newton Telescope at La Palma

The INT dome at La Palma

were simply left where they were. The Castle changed hands several times, and at one time there was a plan to turn it into a country club and golf course!

The situation was saved, rather later than the eleventh hour, when the Queen's University of Toronto bought the Castle to use as a conference centre, and a separate organization was authorized to rescue the telescopes. As yet only one main refractor has been brought back into use, but a science centre has been established, and the outlook seems reasonably bright. Meantime, there was energetic activity on La Palma.

The reconstructed INT was soon in use, and in 1987 it was joined by a still larger reflector, the William Herschel Telescope or WHT. The mirror is 165 inches across, with a total collecting area of 134 square feet, so that of all currently-working single mirrors only those of the Palomar and Russian reflectors are larger. The mirror is of Cervit, and is as accurate as any in existence; it was, incidentally, the last mirror to be made by the Grubb Parsons firm in Newcastle. The total mass of the altazimuth mounting is 190 tons, and the mirror itself weights 16.3 tons. Despite its huge size, the telescope can track a star across any accessible part of the sky without deviating by more than two-tenths of a second of arc.

The WHT has five focal stations where auxiliary instruments can be put. Nobody ever looks through the telescope; where the eyepiece might be expected to be we find ISIS, a double-beam spectrograph.

The Observatory site is imposing by any standards. Los Muchachos is an extinct volcano (at least, one hopes that it is extinct); the name means "the Boys", and the Boys themselves are two rocks on the actual summit. I am sure that there is a local legend about them, but I have never been able to find out what it is.

When I first went there, years ago, the road up the volcano was very rough, if indeed it could be called a road at all; on several occasions my Land Rover was stuck in potholes. Today the road is excellent, and driving up it is no problem at all. From the summit the view is superb, and on a clear day you can even see Mount Teide on Tenerife, where there is another observatory.

La Palma is a Spanish island, but the Observatory is truly international. Telescopes include the 100-inch Nordic reflector, a joint venture by Norway, Sweden, Denmark and Finland; the 40-inch JKT or Jacobus

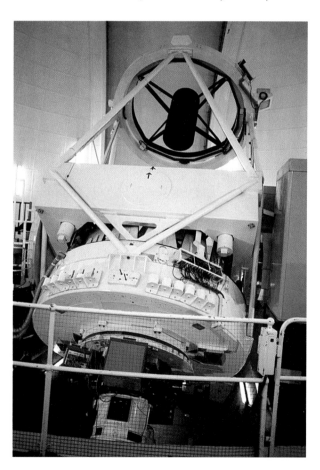

The 100-inch Nordic reflector at La Palma

Kapteyn telescope, named after the Dutch astronomer who discovered the phenomenon of star-streaming; a Swedish solar tower; the Danish Carlsberg transit instrument, which is fully automatic; and a German gamma-ray array, which looks rather like a collection of telephone kiosks. Before long there will also be a giant Italian telescope, named in honour of Galileo.

La Palma is a fine site, with fine telescopes. Official cost-cutting is always a danger, and there have been suggestions that to save money one or more major telescopes may be mothballed, but we can only hope that this does not happen. These instruments are doing splendid work, and are adding greatly to human knowledge. Remember, too, that maintaining a large telescope costs a great deal less than building a nuclear submarine.

Chapter 13
Telescopes of Many Kinds

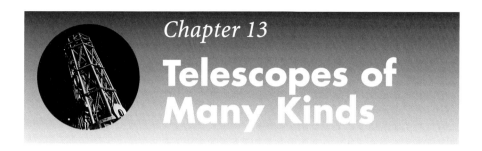

The Multiple-Mirror Telescope at Mount Hopkins, Arizona, in 1992. The six 72-inch mirrors are now being replaced by a single 256-inch mirror

EVEN in modern times there have been some unconventional telescopes, though nothing nearly so outlandish as the Rosse Leviathan. For example, at the Archenhold Observatory in what used to be East Germany there is the Treptow refractor, which has a 27-inch object-glass but a focal length so long – 60 feet – that it looks remarkably like a gun. Amateurs go in for all sorts of designs, such as mounting telescopes on wheelbarrows so that they can be moved around to

escape tree obstruction or light pollution. But there are also large professional telescopes which have departed very much from the norm. One of these was the MMT or Multiple-Mirror Telescope on Mount Hopkins, in Arizona.

Atop Mount Hopkins, one of the peaks of the Santa Rita Mountains, you will find the Whipple Observatory. (I may add that the drive, up a rough, narrow road with a sheer wall on one side and a sheer drop on the other, is not for the faint-hearted.) At first sight the telescope building looks like a square hut. As the telescope inside is moved round, so the whole building turns with it, so that the slit is always in front of the telescope itself.

There used to be not one mirror, but six, each 72 inches in diameter, each weighing only about one-third as much as a conventional mirror of the same aperture. They worked together, so that the light collected was brought to the same focal point, and the result would match a reflector with a single 176-inch mirror – so that when the MMT was set up, in 1979, it was effectively the largest telescope in the world apart from the Hale reflector and the Russian 236-inch. It was owned and used by the University of Arizona, and on the whole it performed well. Naturally it was on an altazimuth, and it looked nothing like a normal telescope.

The main problem was in aligning the six mirrors. The first system involved a laser beam shot through the optical system; an artificial star was generated at the centre of the structure and its image sent out to each of the six mirrors, so that any faults in alignment could be automatically corrected. For a number of reasons this did not work well, and it was abandoned in the early 1980s. Moths tended to fly into the beam at night and interrupt it, shutting down the whole system, and moth recognition with computers was a real difficulty. The laser was then replaced by a more straightforward system, whereby the images at the focal plane of the telescope were digitally recorded.

The MMT had been so designed for two reasons. First, it was cheaper than a single 176-inch reflector would have been. Secondly, it was much easier to make six 72-inch mirrors than one of much greater size. Yet in the mid-1980s the University decided to dismantle the MMT, and substitute a single 256-inch mirror. It was clear that even before the end of the century the MMT in its old form would have dropped from being the third largest telescope in the world down to something like the twentieth, and though this was still a respectable size it would mean that some new branches of frontier research would be unavailable. If the new large mirror were to be mounted, it would have to replace the MMT; there was no other site that could be used.

The 256-inch was to be made by a new method known as the spin-casting technique. The "laboratory" was located underneath the University of Arizona football stadium, and the whole method was due to a leading optical worker, Roger Angel.

The mirror was to be made of borosilicate glass, which is very easy to melt, and to save weight there was to be a honeycomb construction. The procedure was to put the glass into a mould, heat it and then spin it, so that when cooled it would assume the form of a curve. (You can see what is meant if you fill a glass with water and spin it round; the water will heap up at the sides of the glass.) The mirror was cast on 2 April 1992; the rotation speed was 7.4 turns per minute.

The glass took a long time to cool down; it had to fall by one-eighth of a degree through a range of around 5000 degrees. By November the temperature was down to room heat, and all was well. The completed mirror is only $2\frac{1}{2}$ feet thick, and of this about 80 per cent is air, thanks to the honeycomb construction; if you could melt all the glass and make a single mirror out of it, the thickness would be no more than six inches.

Of course the telescope had to be renamed, and indeed a competition was held to decide upon something suitable. All sorts of suggestions were made, but finally the letters MMT were retained; they now stand for Mono-Mirror Telescope. "First light" will not be long delayed.

Astronomers on another Arizonan peak, Mount Graham in the Pinaleños range, were faced with problems of a different kind. This was the outstation of the Vatican Observatory, which had earlier moved its main instruments from Rome into the clearer skies of the Italian countryside. Even this was not satisfactory, and the authorities planned a 70-inch reflector modelled on the lines of the NTT on Mauna Kea; the focal length was to be very short, so that the telescope itself would be remarkably squat and would need only a small dome. The mirror was made by Roger Angel and his team; the mounting was built, and the site on Mount Graham selected. That was where the trouble began.

The 70-inch VATT (Vatican Advanced Technology Telescope) at Mount Graham, Arizona

The Pinaleños are made up of a series of "mountain archipelagos", in which each summit is to a great extent isolated from its surroundings. It so happened that Mount Graham was the habitat of some rare red squirrels. Only a few hundred existed (until a few years earlier they had even been hunted), but they were officially "endangered", and when the Observatory plans were announced there were howls of anguish from the environmentalists. The objections became so vehement that one Arizona biologist who supported the scheme actually received death threats by mail. The arguments went on for years, and it is fair to say that at one stage the astronomers felt that they were being driven rapidly nuts, but eventually a compromise was reached; the Observatory area is fenced in and is very restricted, leaving the rest of the mountain-top to the squirrels. In fact there is no conflict at all, and the numbers of squirrels have increased since the Observatory was built, so that the situation is one of peaceful co-existence. (During my visits to Mount Graham, I have yet to be treated to my first squirrel sighting.)

Also on Mount Graham there is a large submillimetre telescope, a result of collaboration between the University of Arizona and the Max Planck Institute in Germany; it is designed specifically for wavelengths between 0.1 and 1.3 mm, ideal for investigating objects such as the cool, dusty molecular clouds inside which new stars are being formed. This will eventually be joined by the LBT or Large Binocular Telescope, which used to be referred to as the Columbus Telescope before the Politically Correct fanatics objected. (It is interesting to speculate as to how long it will be before these curious people start to protest about terms such as black holes and red giants.) The two 8.4-metre telescopes will be set on the same altazimuth mounting, and will form an interterometer capable of attaining a resolution of 0.005 of a second of arc in the optical range.

Other projects are in hand in various parts of the globe. The Gemini programme involves two 325-inch

The submillimetre telescope on Mount Graham

reflectors, one to be set up on Mauna Kea, north of the equator, and the other in Chile, to the south; working together, they will be able to survey the entire sky – and the mirrors will be so accurate that if either could be enlarged sufficiently to span the Atlantic Ocean, the figure would everywhere be accurate to within one inch. Another giant, the Magellan 6.5-metre reflector, will be sited at Las Campanas.

Last, but by no means least, let us turn to the VLT or Very Large Telescope, sited at Cerro Parañal, south of Antofagasta in the Atacama Desert of northernmost Chile.

I have been to some desolate places, but there is none in my experience to match this part of the Atacama. It takes hours to drive to Parañal from the mining town of Antofagasta, and for most of the way civilization is notable only by its absence. Long before you reach Parañal, over 8000 feet high, all living things have vanished. At the time of my last visit, in January 1997, the daytime temperature would have been over 110 degrees Fahrenheit in the shade – if there had been any shade.

Rainfall occurs about once in a hundred years; the percentage of clear nights is remarkably high, and the air is steady, so that this is probably the best observing site to be found anywhere except for infra-red work, where Mauna Kea is supreme. Before the Observatory building was started nothing lived there; no insects

One of the VLT mirrors

One of the four rotating domes housing the VLT at Cerro Parañal in Chile

(not even scorpions), no birds, no animals. Now that men have arrived you see the occasional desert fox, and a few birds, but little else.

To prepare the site, the top was literally sliced off the mountain. At that stage there was an unwelcome interruption, because a Chilean family claimed ownership, and on one occasion the police actually arrived at the site and stopped all work. Thankfully the problem was resolved, but construction was held up for months.

The VLT is to consist of four 8-metre single mirror telescopes, working together. The four domes are adjacent to each other – not lined up, as originally planned – and the light from all four telescopes can be combined, giving a final resolution of a tiny fraction of a second of arc. Of course the four can be used separately if desired. When the whole installation is complete, the VLT will be very much more powerful than anything previously built.

Go to Parañal, and you cannot fail to be taken aback by the sheer size of the four domes. If the VLT is as good as we have every reason to expect, it will be difficult to quarrel with the claim made by one of the astronomers concerned: "The real future of Earth-based optical astronomy belongs here – on this mountain."

Chapter 14
Telescopes Beyond the Earth

THROUGHOUT this book I have been concentrating upon optical telescopes, designed to study visible light. I have said little about what we may call "invisible astronomy", although it is today of paramount importance. Infra-red radiations were first noted by Sir William Herschel as long ago as 1800, and today we have huge infra-red telescopes, such as the UKIRT and IRTF on Mauna Kea. Radio astronomy began in the early 1930s, with the largely accidental discoveries of Karl Jansky; the 250-foot "dish" at Jodrell Bank came into full operation in 1957, the year which saw the opening of the Space Age.[*] But most radiations outside the visible range are blocked by layers in the Earth's upper air, so that to study them we have to "go aloft".

The first flying observatory of any size was the KAO or Kuiper Airborne Observatory, a converted C.141 cargo plane which carried a 36-inch Cassegrain reflector and could reach a height of 45,000 feet, which is above 85 per cent of the Earth's air and more than 99 per cent of atmospheric water vapour – which, as we have noted, is the main enemy of the infra-red astronomer. The KAO was magnificently successful, and operated for many years without a hitch. It was eventually taken out of service to save money, and not for some time yet will its successor, SOFIA (the Spectroscopic Observatory for Infra-Red Astronomy) be ready. This will carry a 100-inch, equal in size to the

---

[*] And, incidentally, the first *Sky at Night* television programme!

Telescopes Beyond the Earth

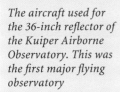

The aircraft used for the 36-inch reflector of the Kuiper Airborne Observatory. This was the first major flying observatory

Inside the KAO

Hooker reflector on Mount Wilson. Until SOFIA is in orbit, the infra-red researchers must wait with as much patience as they can muster.

Yet even these flying telescopes are still well inside the atmosphere, and for perfect conditions we must go higher still – in fact, into space. This was first achieved by means of rockets, and in 1962 instruments on board a rocket were used to identify the first cosmic X-ray source, thereby introducing a whole new branch of astronomical research. Since then, artificial satellites have carried "telescopes" of many kinds. One great

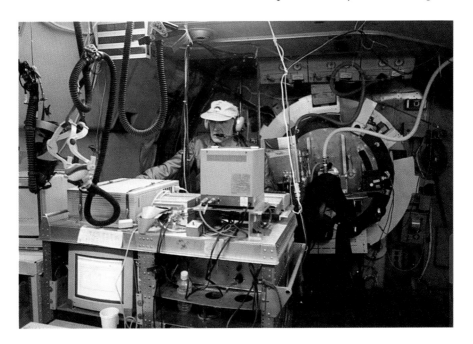

success was with the IUE or International Ultra-violet Explorer, launched in 1977 with a planned operational lifetime of two years; in fact it went on sending back first-class data until 1996, when it was switched off to save money. In the 1970s the American space-station Skylab, manned by three successive three-man crews, carried equipment of all kinds, particularly suited to studies of the Sun, and equipment has also been used on the Russian space-station Mir, still in orbit. The Infra-red Astronomical Satellite IRAS of 1983 carried an infra-red telescope which produced a complete map of the sky in this region of the electromagnetic spectrum; now we have ISO, the Infra-red Space Observatory, which was sent up on 17 November 1995 and put into a path which takes it once round the world in 24 hours, at distances ranging from 43,000 miles to only 620 miles. But when thinking about space astronomy, we always come back to one particular instrument: the Hubble Space Telescope or HST, named in honour of the man who first proved that the universe is in a state of expansion.

The HST took a long time to plan. It was finally launched on 25 April 1990 from the Shuttle *Discovery*, at Cape Canaveral in Florida. Only when it was safely in orbit was it found that there had been an error in shaping the 94-inch mirror, which was too shallow; the total centre-to-edge error was equal to one-fiftieth the thickness of a human hair, but this was enough to give the star images an unacceptable amount of the fault known as spherical aberration. For some time there was a feeling of black despair at the Space Telescope Science Institute at Baltimore, but the situation was rectified by a brilliant repair mission undertaken by NASA astronauts between 2 and 13 December 1993, and by now the telescope is working even better than had been originally expected. A second servicing mission was undertaken in February 1997.

The HST is moving at a height of 375 miles above the Earth, completing one orbit in 95 minutes; the orbital inclination is 28.8 degrees, so that at one time or another the telescope can be seen from most parts of the world. The mirror may be small compared with some of the Earth-based giants, such as the Kecks, but the HST has one overwhelming advantage: it is not troubled by the atmosphere. The figure of the mirror is correct to within two-millionths of an inch, and the wavelength range is from 1200 Ångströms up to one millimetre, so that the telescope is nothing if not versa-

The Space Shuttle Discovery waiting to launch the Hubble Space Telescope into orbit

tile. (One Ångström is equal to one hundred-millionth part of a centimetre.)

The HST has been programmed to point its 25,000-pound mass to within 0.01 of an arc-second of its target object, and to hold it within 0.007 of an arc-second for as long as 24 hours at a time. Improved equipment added during the February 1997 servicing mission means that in theory, the telescope could pinpoint an object the width of a human hair over a range of three miles. One of the most successful of the original instruments was WFPC, the Wide Field and Planetary Camera; a 1997 addition was NIRMOS, the Near Infra-Red and Multi-Object Spectrometer, designed to provide ultra-sharp images of very remote objects.

The HST has now been operating for seven years, though it is only since the 1993 servicing mission that it has realized its full potential. Of its success, there can be no doubt whatsoever. The images of Mars and

The Hubble Space Telescope

the other planets rival those sent back by spacecraft; we can follow the activity on Io, Jupiter's volcanic satellite, and we can make out surface details on Pluto, which has yet to be surveyed from close range. There are pictures of star-formation regions in our Galaxy which can only be described as stunning; we can see details in nebulae, such as that associated with the erratic variable star Eta Carinae; we can resolve globular clusters down to their very centres. Remote galaxies and quasars are shown with almost incredible clarity, and we can see the effects of "gravitational lensing", when the light from a very distant source is distorted as it passes close to an intervening mass. It may well be that in the near future the HST will be able to give direct images of planets moving round other stars.

Inevitably the HST has had its critics. It cost a great deal of money, and it has been claimed that sums of this kind would have been better spent in producing more Earth-based instruments, but the argument does not stand up to examination, because the HST has opened up new avenues of research which would otherwise have remained permanently closed. Its original lifetime was given as 15 years, but the fact that it can be regularly visited and serviced must mean that it will last for much longer than that – always provided that it is not damaged by collision with a piece of "space junk", either natural or artificial, which remains an unpleasant possibility.

For the moment no more orbiting telescopes of the same type are being funded, but when a new project is eventually put in hand we will be able to draw upon the experiences we have had with the HST. It is truly a pioneering instrument, just as Galileo's refractor and Newton's primitive reflector were pioneers in their time.

Epilogue

Into the Future

THIS has necessarily been a very incomplete survey. I have barely mentioned some of the most modern techniques, such as fibre optics and interferometry; all I have tried to do is to give a comprehensive if brief account of the main developments of the past four centuries.

Lastly, what of the future?

Larger and larger telescopes are being made, and – just as importantly – analytical techniques are keeping pace. To all intents and purposes we have left photography behind except for certain special needs, and we are well and truly into the electronic age. During the next few decades we will no doubt be able to set up observatories on the Moon, where there will be no atmospheric interference or light pollution; radio telescopes will be sited on the far side of the Moon, always turned away from the Earth, where the conditions are absolutely "radio quiet". The possibilities are endless, and the new century will indeed be exciting.

We have come a long way since the time of Galileo.

Appendix 1
The History of the Telescope

c.1555   First telescope possibly made, by Leonard Digges.
1608   Telescope constructed by Lippershey.
1609   First telescopic map of the Moon drawn, by Harriot.
1610   Galileo first uses a telescope, and makes a series of discoveries.
1615   Christoph Scheiner observes with the new Keplerian type of refractor.
1637   Founding of the first national observatory (Copenhagen).
1655   Huygens discovers Titan, using an "aerial telescope".
1667   Founding of the Paris Observatory.
1663   First pattern for a reflecting telescope (Gregory).
1668   First reflector made, by Newton (presented to the Royal Society in 1671).
1675   Founding of the Royal Greenwich Observatory.
1723   Construction of the first really good reflector, by Hadley.
1729   Achromatic object-glass described, by Moor Hall.
1759   Achromatic object-glasses made by John Dollond.
1774   First recorded telescopic observations by William Herschel.
1789   Completion of Herschel's 40-foot reflector.
1815   Fraunhofer's first detailed map of the solar spectrum.
1821   Founding of the Cape Observatory.
1824   First telescope to be equatorially mounted and clock-driven (Fraunhofer).
1834–8   First detailed survey of the southern sky (John Herschel).
1838   First measurement of the distance of a star (Bessel).
1844   Founding of the first American observatory, at Harvard.

1845	Completion of the Rosse 72-inch reflector at Birr Castle.
1846	Discovery of Neptune, by Galle and D'Arrest, using a Fraunhofer 9-inch refractor from Berlin. Discovery of Triton by Lassell, with his 24-inch reflector.
1850	First photograph of a star (Vega, from Harvard).
1858	First photograph of a comet (Donati's).
1862	Construction of the 25-inch Newall refractor, by Cooke.
1864	First studies of stellar spectra, by Huggins.
1869	Completion of the Great Melbourne Reflector.
1880	First good image of a gaseous nebula (M.42, by Draper).
1891	Invention of the spectroheliograph, by Hale.
1893	Completion of the Greenwich 28-inch refractor.
1894	Founding of the Lowell Observatory, in Arizona.
1897	Completion of the Yerkes Observatory.
1905	Founding of the Mount Wilson Observatory in California.
1917	Completion of the 100-inch Hooker reflector at Mount Wilson.
1930	Discovery of Pluto, by Tombaugh, using a 13-inch refractor at the Lowell Observatory.
1931	First detection of radio waves from the sky, by Jansky.
1948	Completion of the 200-inch reflector at Palomar.
1957	Work begins with the Lovell 250-foot radio telescope at Jodrell Bank.
1961	Completion of the 210-foot radio telescope at Parkes, Australia.
1967	Completion of the Isaac Newton Telescope at Herstmonceux.
1973	Opening of the Sutherland station of the South African Astronomical Observatories.
1976	Completion of the 236-inch Russian reflector.
1979	Official opening of the observatory on La Palma.
1984	Isaac Newton Telescope transferred to La Palma.
1987	Completion of the William Herschel Telescope on La Palma.
1989	NTT (New Technology Telescope) installed at La Silla.
1990	Launch of the Hubble Space Telescope.
1992	Completion of the first Keck telescope, on Mauna Kea.
1993	Repair mission to the Hubble Space Telescope.
1996	Completion of the Keck II telescope, on Mauna Kea.
1997	Restoration of the 72-inch Rosse telescope at Birr Castle.
1997	Second servicing mission to the Hubble Space Telescope.
1997	Construction of the VLT (Very Large Telescope) at Cerro Parañal in Chile

Appendix 2
Some Great Telescopes

Name	Observatory	Aperture (inches)	Date of completion
Refractors			
Yerkes 40-inch	Yerkes, Williams Bay, Wisconsin, USA	40	1897
Lick 36-inch	Lick, Mount Hamilton, California, USA	36	1888
Meudon 33-inch	Meudon, Paris, France	33	1889
Potsdam Refractor	Potsdam, Germany	31	1899
Lunette Bischoffsheim	Nice, France	29	1886
Treptow Refractor	Archenhold, Treptow, Germany	27	1896
Vienna Refractor	Vienna, Austria	26	1880
26-inch Equatorial	U.S. Naval Observatory, Washington DC, USA	26	1873
Thompson Refractor	Herstmonceux, England	26	1897
Innes Telescope	Johannesburg, South Africa	26	1926
Newall Refractor	Athens, Greece	25	1862
Reflectors			
Keck I	Mauna Kea, Hawaii	387	1991
Keck II	Mauna Kea, Hawaii	387	1996
Bolshoi Teleskop Azimutalnyi	Mount Pastukhov, Russia	236	1975

Name	Observatory	Aperture (inches)	Date of completion
Reflectors (*continued*)			
Hale Telescope	Mount Palomar, California, USA	200	1948
William Herschel Telescope	La Palma, Canary Islands	165	1987
4-metre Telescope	Cerro Tololo, Chile	158	1976
Anglo-Australian Telescope	Siding Spring, NSW, Australia	153	1975
Mayall Reflector	Kitt Peak, Arizona	150	1973
United Kingdom Infra-Red Telescope	Mauna Kea, Hawaii	150	1978
Canada-France-Hawaii Telescope	Mauna Kea, Hawaii	141	1979
3.6-metre Telescope	La Silla, Chile	141	1977
3.5-metre Telescope	Calar Alto, Spain	138	1984
New Technology Telescope	La Silla, Chile	138	1989
Wisconsin–Indiana–Yale–NOAO Telescope	WIYN Observatory, Kitt Peak, Arizona, USA	138	1994
Shane Telescope	Lick, Mount Hamilton, California, USA	120	1959
NASA Infra-Red Facility	Mauna Kea, Hawaii	118	1979
107-inch Telescope	McDonald, Mount Locke, Texas, USA	107	1969
UBC-Laval Telescope	Vancouver, Canada	106	1992
Shajn Telescope	Crimean Observatory, Russia	104	1960
Byurakan Reflector	Byurakan, Mount Aragatz, Armenia	104	1976
Nordic Telescope	La Palma, Canary Islands	101	1989
Isaac Newton Telescope	La Palma, Canary Islands	101	1984
Irénée du Pont Telescope	Las Campanas, Chile	100	1976
Hooker Reflector	Mount Wilson, California	100	1917
Hubble Space Telescope	orbital	94	1990

Some Great Telescopes

Name	Observatory	Aperture (inches)	Date of completion
Schmidt telescopes			
2-metre telescope	Karl Schwarzschild, Tautenberg, Germany	53	1960
Oschin Telescope	Palomar, California, USA	48	1948
United Kingdom Schmidt	Siding Spring, Australia	48	1973
Kiso Schmidt	Kiso, Japan	41	1975
Telescopes in preparation			
Very Large Telescope	Cerro Parañal, Chile	630 (equivalent)	2000
Large Binocular Telescope	Mount Graham, Arizona	465 (equivalent)	1998
Spectroscopic Survey Telescope	McDonald, Texas, USA	433	1997
Suburu Telescope	Mauna Kea, Hawaii	327	1999
Gemini Telescope North	Mauna Kea, Hawaii	319	1999
Gemini Telescope South	Cerro Tololo, Chile	319	2001
Magellan	Las Campanas, Chile	256	1998

Index

AAT (Anglo-Australian Telescope), 72–4, 89
active optics, 78
Adams, J.C., 43
adaptive optics, 78–9
aerial telescopes, 14
Airy, G., 43–4
Almagest, 2
altazimuth mountings, 24, 69–70
Andromeda Spiral (M.31), 58, 64
Angel, Roger, 84, 97
Antofagasta, 100
Arab astronomy, 2, 3
Archenhold Observatory, 95
Aristarchus, 2
Armagh Observatory, 36
Ashurbanipal, Library of, 2
Assyrian astronomy, 2
Atacama Desert (Chile), 77
Athens Observatory, 45
Auzout, A., 15

Baade, W., 63
Bath, City of, 27
Bessel, F.W., 57
Bianchini, map of Venus by, 16
Biela's Comet, 46
Birr Castle, 34–9
Boeddicker, O., 40
Boksenberg, A., 90
Bourne, W., 9
Burleigh, Lord, 9
Byurakan reflector, 70

CCD (Charge-Coupled Device), 64, 90
CFH (Canada-France-Hawaii) Telescope, 84
Campani, telescopes by, 16
Cape Observatory, 74–5
Carlsberg Transit Instrument, 94
Carnegie Foundation, 55, 60
Cassegrain reflectors, 22
Cassini, G.D., 15, 16
Cepheid variables, 57–8, 63–4
Cerro Parañal, 100–1
Cerro Pachón, 80
Cerro Tololo, 77, 79
Challis, J., 44
Charles II, King, 23
Chinese astronomy, 1, 2
Clark, Alvan, 48
Cooke, J., 45, 48
Coonabarabran, 72
Copenhagen Observatory, 15
Copernicus, 4, 11
Crab Nebula, 67
Crimean reflector, 70, 71
61 Cygni, 57

D'Arrest, H., 44
Dee, J., 8
Delhi, observatory at, 3
Delta Cephei, 57–8
Digges, L., 8–9, 25
Digges, T., 8
Dione, 16
Dollond, J., 17
Dollond, P., 17
Doppler effect, 58–9
Dreyer, J.L.E., 40
DuBridge, L., 61

Earth, shape and status of, 1, 2
eclipses, solar, 2
Egyptian astronomy, 1, 2
Enceladus, 31, 32
Encke, J., 44
equatorial mountings, 24
Eta Aquilae, 58
Eta Carinae, 106
Euclid, 8

false colour problem, 13, 14, 17, 19
Feldhausen, 33
Flagstaff (Arizona), 49
Flamsteed, J., 23–4
  – star catalogue by, 23
Foucault, J., 22
Fraunhofer, J., 35, 43, 53

galaxies, recession of, 58
Galileo, 7, 10–12, 25, 53, 108
Galle, J., 44
gamma-ray telescope, La Palma, 94
Gascoigne, W., 25
Gemini telescopes, 99–100
George II, King, 26
George III, King, 29
Greek astronomy, 2
Greenwich Observatory, 23, 24, 43–5, 88–92
Gregorian telescopes, 19, 20
Gregory, J., 19, 20
Grosseteste, R., 8
Grubb, T., 46
  – 41in refractor, 52
Grubb Parsons Ltd., 92

HST (Hubble Space Telescope), 104–7
Hadley, J., 25
Hale, G.E., 54–7, 60–1
Hale Telescope, *see* Palomar 200in reflector
Hale Pohaku, 83

Hall, C.M., 17
Halley, E., 17, 24
Halley's Comet, 68
Hawaii, 72
Henry brothers, 48
Harriot, T., 9
Harrison, T., 23
Herschel, A., 27
Herschel, Caroline, 27
Herschel, Jacob, 26
Herschel, John, 33, 74–5
Herschel, William, 25, 26–33, 43, 102
  – 40ft reflector by, 30–3
Herschelian telescopes, 31
Herstmonceux Castle, 88–92
Hevelius, 14, 23
Hipparchus, 2
Hooker reflector, 56–9, 103
Hubble, E., 58, 63, 64
Hubble Constant, 58–9
Humason, M., 58
Huygens, C., 15, 16, 25
Hven, 5

INT (Isaac Newton Telescope), 88–90
IRTF (Infra-Red Telescope Facility), 84, 102
ISO (Infra-red Space Observatory), 104
IUE (International Ultra-violet Explorer), 104
Iapetus, 16
infra-red radiation, 83
Innes refractor, 50, 75, 102
Io, 106

JCMT (James Clerk Maxwell Telescope), 84
JKT (Jacobus Kapteyn Telescope), 93
Jansen, Z., 9
Jansky, K., 65, 102
Johannesburg, city, 50
Jupiter, 10–11, 39, 57
  – satellites, observed by Galileo, 10–11

KAO (Kuiper Airborne Observatory), 102
Kapteyn, J., 94
Karl-Schwarzschild Observatory, Schmidt telescope at, 65
Keck Telescopes, 84–7
Kepler, J., 5, 6, 10
Kew Observatory, 29, 30
Kitt Peak Observatory, 76
Kuiper, G., 82, 83

LBT (Large Binocular Telescope), 99
La Palma, 89
La Serena, 77
La Silla Observatory, 77–9
Las Campanas Observatory, 77, 79–80
Lassell, W., 45, 46
latitude, finding, 22–3
Le Verrier, U.J.J., 43–4
Leonardo da Vinci, 8
Lick refractor, 48–9
light, refraction of, 13
Lippershey, H., 7, 8, 9
longitude, finding, 22–3
Los Muchachos, 89
Lovell Telescope, 66–7
Lowell, P., 49
Lowell refractor, 48–9, 50, 58
Lower, W., 10

MMT (Multiple/Mono Mirror Telescope), 95–7
Magellan reflector, 110
Magellanic Clouds, 71
Malin, D., 74
Marius, S., 11
Mars, 5, 15, 39, 49, 105
Maskelyne, N., 29
Mauna Kea, 82–4
Mauna Loa, 83
Mayall reflector, 76
McMath Solar Telescope, 76–7
Melbourne reflector, 46–8
Mesopotamian astronomy, 1
Messier, C., 38, 57
Meudon refractor, 48
Milky Way, seen by Galileo, 10
  – radio waves from, 65
Mimas, 31, 32
Mir, 104
Moon, 1, 7, 10, 14, 39, 40, 108
Mount Graham, 97
Mount Hamilton, 48–9
Mount Hopkins, 96
Mount John, 80–1
Mount Pastukhov, 69
Mount Palomar, 55, 60–64
  – 200in reflector, 60–4, 67–8, 92
Mount Wilson, 54, 55, 56–9

NTT (New Technology Telescope), 78–9
nebulae, 38–9
Nelson, J., 84–6
Neptune, discovery of, 44
Newall Telescope, 45
Newton, Isaac, 9, 19, 21, 22–4
  – telescope by, 20–1
Newtonian reflectors, 21
Nordic Reflector, 93
Northumberland Telescope, 44–5

object-glasses, 7
  – achromatic, 17
Omega Centauri, 72
Orion Nebula, 27, 38, 46
Oschen Schmidt (Palomar), 65

Palomar reflector, 60–4, 67–8, 92
parallax, 57
Paris Observatory, 15–7
Paris 49in refractor, 51–2
Parkes Observatory, 67
Pechan, J., 8
photography, 33, 53, 74
Pinaleños Range, 97–8
precession, 2
Pluto, 106
Ptolemaic theory, 2, 4
Ptolemy, 2, 4
Pulkovo Observatory, 70

quasars, 67–8
Queen's University, 92

Radcliffe Observatory, 75
radio astronomy, 65–7, 108
reflector, principle of, 7
refractor, principle of, 7
Rhea, 16
Ritchey, G.W., 47, 55
Robinson, Romney, 36, 37, 39
Ronan, C.A., 8, 9
Rosse, 3rd Earl of, 34–9, 46, 57
  – 36in telescope by, 35
  – 72in telescope by, 36–42
Rosse, 4th Earl of, 40
Rosse, 6th Earl of, 41
Royal Greenwich Observatory, *see* Greenwich
Russian 236in reflector, 69–70, 71, 86, 92

SOFIA (Spectroscopic Observatory for Infra-Red Astronomy), 102–3
Samarkand, observatory at, 3
Santa Rita Mountains, 96
Satterthwaite, G., 9
Saturn, 10, 15
Scheiner, C., 10
Schröter, J., 30
Schmidt cameras, 64–5
Schmidt, B., 64
Schmidt, M., 67
Short, J., 21, 25
Siding Spring Observatory, 72–6
Sirius, heliacal rising of, 2
Skylab, 104
Slipher, V.M., 58
Slough, town, 30
Snow Telescope, 55
South Africa, astronomy in, 74–5
South, J., 35
South Pole, observatory at, 81
Southern Cross, 72
spectroheliograph, invention of, 54
spectroscopy, 20, 53–4
speculum metal, 22
spiral galaxies, 38
squirrels, endangered, 98
stars, distances of, 57
Sun, status of, 1
  – distance of, 25
  – photography of, 53
sunspots, 11, 53
Suburu telescope, 84, 85
Sutherland Observatory, 75

Teide (Tenerife), observatory at, 93
telescope mountings, 24
Tethys, 16
Thales, 2
Titan, 15
Treptow telescope, 95
Triton, 45
47 Tucanae, 72
Tycho Brahe, 4, 5, 23
Tychonic system, 5

UKIRT (United Kingdom Infra-Red Telescope), 83, 102
UKS (United Kingdom Schmidt), 74
Ulugh Beigh, 3
Uranus, 28, 43

VATT (Vatican Advanced Technology Telescope), 98
VLT (Very Large Telescope), 100–1
Vatican Observatory, 97
Vega, 57
Venus, phases of, 11
  – dawn by Bianchini, 16
  – transit of, 29
Vienna refractor, 46

WHT (William Herschel Telescope), 92
Warrumbungle Mountains, 72
Whipple Observatory, 96
Witelo, 8
Wren, Christopher, 23

X-rays, 103

Yerkes, C.T., 49, 54
Yerkes refractor, 48–9, 54

❝ If you are reading this book, or if you are a regular watcher of *The Sky at Night*, then it is very likely that you are interested in astronomy in quite a serious way.

The Practical Astronomy series – of which I am Series Editor – has been put together specially for amateur astronomers who, while not yet experts, want to learn more. The last few pages of this book are devoted to giving brief information about some of these books. They are all books that I have been personally involved with, and I can heartily recommend them.

They are, of course, available in all good booksellers (who can order them even if they are not in stock), but if you have any difficulties you can contact the publishers direct, by telephoning +44 (0)1483 418822 (in the UK and Europe), +1 /212/4 60 15 00 (in the USA), or by emailing postmaster@svl.co.uk

I hope you will continue to enjoy astronomy as much as I do! ❞

Patrick Moore

THE OBSERVATIONAL AMATEUR ASTRONOMER

Dr. Patrick Moore CBE (editor)

This book deals entirely with serious observing, and pulls together chapters by noted professional and amateur observers, each covering a different category of object, from the Moon to deep-space. It explains how to use a small telescope for serious observing, whether from backyard, field location, or university. The ideal companion to *The Modern Amateur Astronomer*.

288 pages, over 75 illustrations and photographs. Softcover.
ISBN 3-540-19899-7

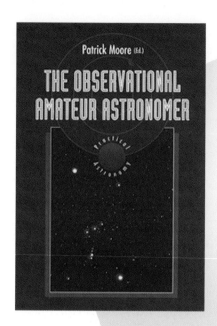

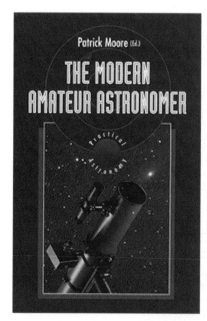

THE MODERN AMATEUR ASTRONOMER

Dr. Patrick Moore CBE (editor)

Along with its companion book, *The Observational Amateur Astronomer,* this provides a comprehensive but affordable guide for amateur astronomers who want to do more than just star-gaze. All aspects of astronomy are covered, from optical principles to astrophotography. Each chapter is written by a well-known professional or amateur astronomer, and care has been taken to put in plenty of information on each topic, but not too much. Patrick Moore has edited the book overall into his easy, readable style.

176 pages, 77 illustrations and photographs. Softcover.
ISBN 3-540-19900-4

THE ART AND SCIENCE OF CCD ASTRONOMY

David Ratledge (editor)

Charge-coupled Devices (CCDs) have revolutionised astronomy. Even affordable CCD cameras can be ten times as sensitive as photographic film, and they deliver a digitised image that is easy to enhance using a personal computer.

David Ratledge has brought together contributions from twelve leading amateurs from around the world, people who are routinely producing astronomical images of a quality that rivals those of professional observatories only of 10 years ago.

These experts describe their techniques and solutions, and offer essential tips and advice for anyone who is choosing or using a CCD camera. There's a Colour Gallery at the back of the book, so you can see just what can be done with amateur equipment!

192 pages,
175 illustrations
and photographs,
70 in colour.
Softcover.
ISBN 3-540-76103-9

TELESCOPES AND TECHNIQUES
An Introduction to Practical Astronomy

Dr. C.R. Kitchin

Chris Kitchin has produced an easy-to-read book explaining how to use a small telescope and find your way around the sky. Covering all the basic topics – telescopes, optics, positions and motion, observing, and instruments – the thing that makes this book "different" is its impeccable scientific background: it has been developed from the author's long experience in teaching undergraduates.

216 pages, 113 illustrations. Softcover.
ISBN 3-540-19898-9

SMALL ASTRONOMICAL OBSERVATORIES

Dr. Patrick Moore CBE (editor)

The advantages of a permanent site for a small astronomical telescope are very clear. It is more comfortable, much quicker to bring into use, and allows you to spend your time observing and not setting up equipment.

But just how difficult is it to build a small observatory? And how much will it cost?

In *Small Astronomical Observatories,* Patrick Moore has collected descriptions of all manner of projects, from several different countries. Each chapter describes a different design – everything from run-off sheds to elaborate rotating domes. Their builders' extraordinary innovations and improvisations make fascinating, instructive, and often amusing reading.

Here is a book that contains a wealth of practical advice for anyone thinking (or dreaming) of building their own observatory.

240 pages,
128 illustrations
and photographs.
Softcover.
ISBN 3-540-19913-6

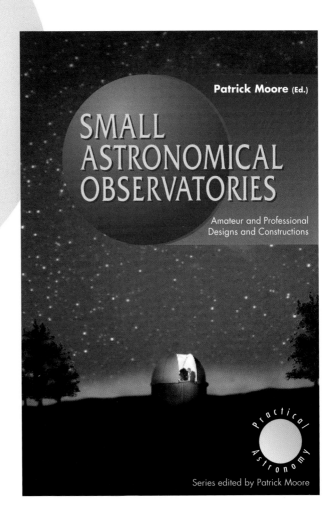

UNVEILING THE UNIVERSE

J.E. van Zyl
(Foreword by Patrick Moore)

Start here if you want to *really* learn about astronomy. Accessible enough to be used by the relative newcomer to the subject – and with lots of illustrations and colour photographs – this is perhaps one of the most complete introductory astronomy books ever published.

You can begin by reading the book for its lucid, detailed and authoritative text.

But as you progress you will want to study the science of astronomy in more detail – subjects like the mass-luminosity law, Planck's laws, and the period-luminosity law – and you will discover that here too, the author has produced a work of unparalleled clarity and completeness.

This will become one of your major reference books for years to come.

340 pages, 383 illustrations and photographs, including 52 in colour.
Hardcover.
ISBN 3-540-76023-7

Not in the Practical Astronomy series, but a "must" for every amateur astronomer...